AF768604

CONTENTS

WITNESSES

TUESDAY, JANUARY 24, 2006

ALPHABETICAL LIST OF WITNESSES

APPENDIX

PREPARING FOR A CATASTROPHE: THE HURRICANE PAM EXERCISE

TUESDAY, JANUARY 24, 2006

U.S. SENATE,
COMMITTEE ON HOMELAND SECURITY AND
GOVERNMENTAL AFFAIRS,
Washington, DC.

The Committee met, pursuant to notice, at 10 a.m., in room SD–342, Dirksen Senate Office Building, Hon. Susan M. Collins, Chairman of the Committee, presiding.

Present: Senators Collins, Lieberman, Levin, and Carper.

OPENING STATEMENT OF CHAIRMAN COLLINS

Chairman COLLINS. The Committee will come to order. Good morning.

Today, the Committee on Homeland Security continues our investigation into Hurricane Katrina. Over the last 4 months, we have conducted 10 hearings on major aspects of the causes and management of this disaster, including a field visit exactly 1 week ago to the Mississippi Gulf Coast and to New Orleans. Our staff has conducted more than 200 interviews and has reviewed more than 300,000 pages of documents.

Now, we are about to enter the final phase of our work. Hurricane Katrina proved to be one of the deadliest and certainly the most costly natural disaster in America's history. If our Nation cannot give a good account of our ability to manage such a predicted, known, and trackable event as a hurricane, we must surely question our preparedness for dealing with a stealthier, more sinister terrorist attack.

Therefore, based on all we have learned from our previous hearings, interviews, and document review, this Committee will undertake over the next 3 weeks a series of hearings to cover the most troubling aspects of the response to Katrina as a prelude to drafting our final report.

The focus of today's hearing is the simulation called Hurricane Pam, a federally funded exercise to plan for a catastrophic hurricane in Southeast Louisiana. We will examine both the lessons learned and the lessons that with such terrible consequences went unlearned. This hearing is intended to shed light on the following issues: How did Hurricane Pam come about? Who took the initiative to promote it? What does its history say about the state of emergency preparedness in Louisiana prior to Katrina? What roadblocks had to be overcome to get Federal funding for the exercise in both President Clinton's and President Bush's Administrations?

Do these roadblocks raise concerns about government priorities in improving emergency preparedness? What was the scope of Hurricane Pam, including assumptions about the specific planning scenarios? How did pre-storm evacuation come to be largely excluded from the exercise? Did Hurricane Pam create the impression within FEMA that Louisiana had evacuation under control? Why was the plan not completed? How did the failure to complete the plan affect its usefulness in Katrina? What aspects of the draft Pam plan were used in responding to Katrina? What aspects could have been used but were not?

The Hurricane Pam exercise was conducted in Louisiana by FEMA from July 16 through 23, 2004. It brought together as many as 300 local, State, and Federal emergency response officials. This fictional storm was designed as a slow-moving Category 3 hurricane that had sustained winds of 120 miles per hour at landfall. It caused as much as 10 to 20 feet of flooding throughout most of New Orleans and the surrounding parishes as the result of heavy rain and a storm surge that overtopped the levees. Pam's mock damage spread over 13 Louisiana parishes and was extensive. In the scenario, utilities were knocked out and chemical plants were flooded. The human cost under the scenario was staggering. More than a million people evacuated, 175,000 were injured, 200,000 became sick, and as many as 60,000 lives were lost.

As a dry run for the real thing, Pam should have been a wake-up call that could not be ignored. Instead, it seems that a more appropriate name for Pam would have been Cassandra, the mythical prophet who warned of disasters but whom no one really believed. In many ways, the hypothetical problems identified in Pam predict with eerie accuracy the all-too-real problems of Katrina—overcrowded shelters undersupplied with food, water, and other essentials; blocked highways with thousands of people trapped in flooded areas; hospitals swamped with victims and running out of fuel for their emergency generators. The list goes on and on.

The history of Pam dates back to 1998, when New Orleans experienced a near-miss from another hurricane. In the fall of 1999, local, State, and Federal officials met to discuss their concerns about the adequacy of plans to respond to a direct hit on the city. The State of Louisiana followed up with a written request to FEMA in August 2000 for a planning exercise. But delay followed delay. Then FEMA reduced the funding allocation so the scope of the exercise had to be scaled back. In reaction, the State agency chose to exclude the critical issue of pre-landfall evacuation and the possibility that the levees could be breached rather than merely overtopped.

The Pam exercise that finally commenced in July 2004 was supposed to be just the first installment of an ongoing process. A follow-up session scheduled for September 2004 was postponed and critical workshops were not reconvened until late July 2005, with the result being that no additional planning documents were generated before they were so urgently needed.

Instead, Pam became Katrina. The simulation became reality. And optimism became the awful truth. We were not prepared.

There are instances in which the Pam exercise did improve the response to Katrina. For example, the Louisiana National Guard incorporated lessons regarding the staging and distribution of such essential commodities as food and water. The State Department of Health and Hospitals adopted concepts developed in Pam on how to evaluate individuals saved through search and rescue efforts.

Our witnesses today represent a wide range of entities involved in the Hurricane Pam exercise. I'm very interested in hearing their frank views on the questions that I raised earlier.

An evaluation of the Pam simulation is important for at least two reasons. First, the stated purpose of the Hurricane Pam exercise was not fulfilled when it counted, with catastrophic consequences. Second, throughout our Nation, local, State, and Federal emergency response agencies engage in a great many training exercises at considerable expense in anticipation of a wide range of natural and manmade disasters. We must use and learn from the experience of Pam and Katrina to close the gap between planning and execution so that we are better prepared the next time simulation becomes reality.

Senator Lieberman.

OPENING STATEMENT OF SENATOR LIEBERMAN

Senator LIEBERMAN. Thanks very much, Madam Chairman, for convening this 11th hearing in our investigation of how the government prepared for and responded to Hurricane Katrina, and as you said, this one begins a 3-week series of hearings in which we have the opportunity to make public a lot of the hard work that our staffs have done in investigating what happened.

The title of today's hearing is, "Preparing for a Catastrophe: The Hurricane Pam Exercise." Preparing for a catastrophe—the phrase makes a mournful sound when said against the backdrop of the misery and destruction the world saw on television last year and that Members of this Committee still saw last week when we visited the Gulf Coast and held a hearing in Mississippi. The plain facts are that Katrina was a very powerful storm, but it would have caused much less misery and destruction had we prepared for it better.

This enlargement from the New Orleans *Times-Picayune,*[1] August 30, the day after Hurricane Katrina hit landfall, really tells it all. The big banner word is "Catastrophic," which it was. But in smaller red print at the top, over the masthead, it also tells it all. "Katrina: The Storm We've Always Feared"—the storm people in the Gulf Coast had always feared, the storm people knew would hit one day, the storm they actually practiced for in the Hurricane Pam exercise that is the topic of today's hearing.

In the 10 Committee hearings on Katrina we have already held, in our staff interviews of more than 200 witnesses, in our review of tens of thousands of documents, we have already learned enough to be not just disappointed, but truly infuriated by the poor performance of all levels of government in preparing for and responding to Hurricane Katrina, and these conclusions, amplified as I am confident they will be over the coming weeks, should compel us to

---

[1] Exhibit E submitted for the Record by Senator Lieberman appears in the Appendix on page 94.

achieve top-to-bottom reform of the way we prepare for and respond to disasters.

Katrina was not just predictable, it was predicted over and over again. As the FEMA Coordinator for the Hurricane Pam exercise told our investigative staff last Friday, Katrina was a "replication" of Pam and Pam itself was staged in response to the flooding in Biloxi, Mississippi, in 1998 caused by Hurricane Georges that made State and local officials of the Gulf Coast realize they could be overwhelmed if and when the "big one" hit. The Hurricane Pam exercise in the spring and summer of 2004 actually and eerily predicted the emergency response crises and the devastation that occurred last August and September.

Today, we are going to hear from four witnesses who participated in the Hurricane Pam exercise who will tell us that the problems we saw last August and September were known long before Katrina hit the Gulf Coast, even long before Pam predicted them. The fictional hurricane of the Pam exercise was a slow-moving Category 3 hurricane, quite similar to Katrina except that in the fictional exercise, Pam hit New Orleans directly, and as we know, thank God, Katrina blew about 15 miles to the east of the city. Had Katrina hit New Orleans head-on as the Pam exercise predicted for Pam, 67,000 deaths would have resulted. That is what the Pam exercise projected. That gives us an idea of how much more catastrophic Katrina could have been and therefore how much more urgent disaster preparation should have been.

The Pam exercise also put State and local governments and FEMA and its parent, the Department of Homeland Security, on notice that the people of New Orleans would experience exactly the problems that we all witnessed last August that Senator Collins has spoken to. The Pam exercise also predicted widespread flooding throughout New Orleans, hospitals and nursing homes underwater, hundreds of thousands of people displaced, and local first responders incapacitated. In this regard, Pam gave DHS and FEMA explicit notice that State and local governments would be overwhelmed when New Orleans got hit with a catastrophic hurricane and that comprehensive Federal assistance would, therefore, be critically and urgently necessary.

But despite these warnings from Pam, preparations for Katrina were shockingly poor. Two to 3 days before Katrina hit, it became clear that it would be catastrophic. In fact, as Katrina approached the Gulf Coast 2 days before landfall, Saturday, August 27, our staff has obtained a document which shows that FEMA issued a briefing at 9 a.m. on that Saturday morning before the Monday of landfall which declared that the Pam "exercise projection is exceeded by Hurricane Katrina real-life impacts." The failure to heed the fictional Pam's many warnings compounded the tragedy when Katrina hit in real time and full fury. That is the sad story that our Committee's hearings will tell in detail in the 3 weeks ahead.

Before closing and as we embark on this stage of the investigation, I feel compelled to say a few words about the conduct of the investigation. First, I want to thank Chairman Collins and her staff for working with me and my staff to conduct an aggressive and thoroughly bipartisan investigation. This has become our norm on this Committee, but I don't want the Chairman to think that

I take it for granted. We have worked together as all investigative committees in this Congress should, without partisan division and with a shared view that our goal is to uncover what happened with respect to Hurricane Katrina so that we can make sure our government is much better prepared the next time disaster strikes.

Unfortunately, though, I cannot give the same high marks to the Executive Branch for its response to our investigation, and the problems begin at the White House, where there has been a near total lack of cooperation that has made it impossible, in my opinion, for us to do the thorough investigation we have a responsibility to do. Why does this matter? Well, here is an example.

The Committee has found evidence that we will describe in the hearings ahead that beginning on Friday before the Monday of landfall, there are explicit statements in e-mails by high-ranking officials at FEMA which show they understood the severity of the storm that was coming—Friday, the document I quoted earlier on Saturday morning, and then on the evening before Katrina made landfall, that Sunday, the Department of Homeland Security circulated to Federal agencies sitting in the Homeland Security Operations Center a report that the storm had at that time been upgraded to Category 5 and that "any storm rated Category 4 or greater will likely lead to severe flooding and/or levee breaching. This could leave the New Orleans metro area submerged for weeks or months."

Among the offices receiving that memo was the White House Situation Room, which received it at 1:47 a.m. on Monday, August 29, several hours before Katrina made landfall. What happened to that report and the other awareness that FEMA officials and others at DHS had of the severity of the coming storm? Why was the President of the United States left so uninformed that he said 3 days later, "I don't think anyone anticipated the breach of the levees."

At this point, we cannot answer that critical question because the White House has produced just a very small portion of the documents we requested. In addition, they have opposed efforts to interview White House personnel and they have hindered our ability to obtain information from other Federal agencies regarding White House actions in response to Katrina. I have been told by my staff that almost every question that has been asked Federal agency witnesses regarding conversations with or involvement of the White House has been met with a response that they could not answer on direction of the White House. There has been no assertion of executive privilege; just a refusal to answer questions.

Indeed, as recently as yesterday in his staff interview, that is, interview with our staff, former FEMA Director Michael Brown's agency lawyers advised him not to say whether he spoke to the President or the Vice President or comment on the substance of conversations he had with any other high-level White House officials. This assertion of a kind of virtual immunity of the White House from this inquiry has obviously frustrated our Committee's ability to learn and tell the full story of Katrina. In my opinion, it is unacceptable.

While some agencies like FEMA, and I want to stress this, have been very cooperative, other executive agencies, including the Department of Justice and the Department of Health and Human

Services, have essentially ignored our document and information requests for months and to this day have produced much less than half the information we asked for. HHS has produced not a single requested witness for an interview, and the Department of Homeland Security, which is at the center of our investigation because it has overall responsibility for national disaster preparedness and response, including in Katrina, has produced too little, too late. Repeated requests for critical witnesses and documents have been ignored or delayed.

My staff on this investigation believes that the Department of Homeland Security has engaged in a strategy of slow walking our investigation in the hope that we would run out of time to follow the investigation's natural progression to where it leads. I hope they are wrong, but at this time, I cannot disagree.

Madam Chairman, I do want to thank you publicly for your continuing efforts to elicit more cooperation from the Administration. I hope the Committee will continue to pursue all these unanswered questions asked of the Executive Branch until we have the information we need to answer the questions that must be answered. In the meantime, because hurricane season begins again in June and the threat of terrorist attacks persists, and because our staffs together, notwithstanding the difficulties I have described, have done some excellent investigative work, these hearings are ready to go forward and must go forward and the Committee's report must be written as soon as possible to help American Government be better prepared to protect America's people from disasters that history tells us will come, disasters that are natural or unnatural.

In that spirit and with thanks to you, I look forward to today's witnesses and those that follow in the 3 weeks ahead. Thank you.

Chairman COLLINS. Our four witnesses today represent State, local, and Federal Government, as well as the private sector entities most involved in the development of the Hurricane Pam exercise. Wayne Fairley is the Response Branch Chief for FEMA in Region VI, I believe it is. As such, he oversees regional operations, logistics, and planning. He has served with FEMA for 24 years. Before that, he served in the Louisiana State Government. He was involved in discussions of a federally -funded catastrophic plan for Southeastern Louisiana since 1999 and was involved in designing, planning, and the exercising of Hurricane Pam as a member of the steering committee.

Sean Fontenot was in charge of the planning at the Louisiana Office of Homeland Security and Emergency Preparedness in the late 1990s when the concept of a federally unded exercise to plan for a catastrophic hurricane was first discussed. In May 2005, he joined the Innovative Emergency Management Company as an emergency planner.

Jesse St. Amant is the Director of the Plaquemines Parish Office of Homeland Security and Emergency Preparedness. He is also the President of the Southeastern Louisiana Hurricane Task Force. He participated in the Hurricane Pam exercises and the follow-up meetings in 2005.

Madhu Beriwal is President and CEO of Innovative Emergency Management, Incorporated. IEM is a Baton Rouge-based research company that works with emergency managers in the public and private sectors to develop and improve their emergency preparation and response capabilities. IEM led a team of three firms that developed the Hurricane Pam scenario under contract with FEMA.

I want to welcome all of you to the Committee today. We very much appreciate the cooperation you have already given us, and we will begin with Mr. Fairley.

TESTIMONY OF WAYNE FAIRLEY,[1] CHIEF, RESPONSE OPERATIONS BRANCH, RESPONSE AND RECOVERY DIVISION, REGION VI, FEDERAL EMERGENCY MANAGEMENT ADMINISTRATION, DENTON, TEXAS

Mr. FAIRLEY. Good morning, Madam Chairman and Members of the Committee. I am honored to appear before you today to discuss this subject and to further any discussions I have had with your various staff over the past week and to answer any questions you may have.

To start off with, I believe it is best to provide a little historical background on the Hurricane Pam exercise. As I recall FEMA's goal based on the 2003 Catastrophic Initiative was to identify areas of the country that could be vulnerable to catastrophic disasters and in cooperation with the relevant State and local governments to examine projected damages and effects associated with catastrophic disasters, confirm current disaster response capabilities, identify anticipated response shortfalls, and to initiate comprehensive planning strategies to address these shortfalls. Products developed under the Catastrophic Planning Initiative were envisioned to include incident-specific response plans for pre-selected geographic regions and disasters, planning templates that could be applied to other areas, and new response contingencies.

In late March 2004, FEMA headquarters notified FEMA Region VI that the State of Louisiana had been funded for a catastrophic hurricane plan. Thirteen Southeastern Louisiana parishes, including the City of New Orleans, were selected as the initial geographic focus for FEMA's Catastrophic Planning Initiative because of their vulnerability to hurricane disasters. This resulted in the Southeast Louisiana Catastrophic Hurricane Planning Project. The initial concept was to have a draft plan by the end of July 2004.

The Southeast Louisiana Catastrophic Hurricane Planning Project was designed to bring together responders and decisionmakers from all levels of government and the American Red Cross to begin analyzing and addressing the overwhelming operational complexities that would be involved in responding to a catastrophic hurricane striking Southeast Louisiana. Accepting the fact that only limited funding and time were available, topic-specific planning workshops using a catastrophic hurricane scenario called Hurricane Pam to frame these discussions were selected as the best approach for identifying and qualifying the scale of requirements needed to build a plan for responding to a catastrophic hurricane. The results were intended to reveal to the Louisiana Office of Homeland Security and Emergency Preparedness and FEMA the shortfalls in existing plans and to begin developing additional plans for catastrophic hurricane response.

---

[1] The prepared statement of Mr. Fairley appears in the Appendix on page 41.

Existing plans, strategies, policies, and capabilities were reviewed by LOHSEP before the first workshop. As preplanning for the first workshop conducted in July 2004, the Louisiana Office of Homeland Security and Emergency Preparedness and Federal representatives identified a list of planning topics based on those provided by the State of Louisiana as the most urgent or complex topics needing discussion, to include hurricane pre-landfall issues, search and rescue, temporary medical care, sheltering, temporary housing, schools, and debris.

During the first workshop, participants were presented with a catastrophic hypothetical Hurricane Pam disaster scenario to frame discussions and then divided into breakout groups by responsibilities and topic for detailed discussions. The breakout groups identified operational concerns in each of the topical areas, addressed issues, and drafted plans for dealing with the identified concerns. To address other urgent subtopics that emerged during the discussions, additional breakout groups were established. The following additional subtopics were discussed: Access control and reentry; billeting of Federal response workers; distribution of ice, water, and power; donations management; external affairs; hazardous materials; transition from rescue to temporary housing; and unwatering of levee enclosed areas.

It became clear after the first workshop that a series of workshop cycles would be needed to address the full range of complex response and recovery concerns associated with this type of catastrophic event. Additional workshops were held in November 2004, July 2005, and August 2005 to provide further input for topics. Topics selected for further discussion during the subsequent workshops included the following. In November, sheltering, temporary housing, and temporary medical care. In July, transportation, staging, and distribution of critical resources and temporary housing. And in August, temporary medical care.

The goal of the Southeast Louisiana Catastrophic Hurricane Planning Project was to begin addressing immediate, intermediate, and long-term needs; create plans immediately usable by planners and responders in the field; and to seed the eventual development of a comprehensive and systematic operational plan The ultimate goal is for the concepts identified in the Southeast Louisiana Catastrophic Hurricane Planning Project to be integrated into a final catastrophic plan. The project did not result in a catastrophic planning document per se, but rather a framework for developing such a plan.

My participation in the process included working with the Louisiana Office of Homeland Security and Emergency Preparedness counterpart as a member of the steering committee. That involved project management, workshop design and participation, budgeting, and headquarters and contractor interface. At the workshops, this included monitoring the workshop sessions; providing FEMA law, regulation, and policy information; dispute resolution; and overall directional guidance in meeting our workshop objectives.

Participation included the Louisiana Office of Homeland Security and Emergency Preparedness, State emergency support functions, local emergency management staff from the 13 Southeast Louisiana parishes, FEMA Region VI, FEMA headquarters, FEMA emergency support functions, other Federal agencies as requested, and private industry partners.

Areas of responsibility were assigned in the workshops according to existing State and Federal laws, regulations, policies, procedures, and plans. No planning effort was made to recreate or modify any existing authority. Directed or institutional agency authority on any given subject area was only reviewed and used as guidance by the planning session participants. However, participants were able to comment and provide opinions on existing State and Federal laws, regulations, policies, procedures, and plans and the possible need for future changes. Two such State plans included the Louisiana Hurricane Evacuation Plan and the Louisiana Sheltering Plan.

These planning sessions laid the groundwork for future detailed subject plans. They identified the primary areas of concern by the local, State, and Federal agencies and began the process of identifying who would address these areas and how they would be addressed. These sessions brought together persons responsible for the implementation of emergency management from all levels of government and helped lay a groundwork of cooperation that had never existed before.

Future intentions were to include continued subject-specific sessions. Some topics were to be expanded. Some topics would be added. Some topics would only be maintained with updated data. It was our hope that the plan would not end or become stagnant but would continue to be a fresh and growing plan that included new data and innovative ideas. It was also hoped that the new-formed working spirit between local, State, Federal, and private industry would continue to grow and lead to a concept of "ours" versus yours or mine.

Although the catastrophic planning process has been interrupted by the impacts of Hurricanes Katrina and Rita, the workshops and planning process—knowledge of inter-jurisdictional relationships and capabilities, identification of issues, and rudimentary concepts for handling the consequences—have been quite beneficial to all involved in the hurricane response.

I know that this Committee and others are concerned about what occurred as a result of Hurricane Katrina in Louisiana, and I want to assure this Committee that all of my fellow employees at FEMA are also concerned. I want to assist this Committee in any way I can in ensuring that what occurred never happens again. I want to thank the Members of this Committee for their past support of FEMA and appreciate the opportunity to testify before you today.

Chairman COLLINS. Thank you, Mr. Fairley. Mr. Fontenot.

TESTIMONY OF SEAN R. FONTENOT,[1] FORMER CHIEF, PLANNING DIVISION, FORMER CHIEF, PREPAREDNESS DIVISION, LOUISIANA OFFICE OF HOMELAND SECURITY AND EMERGENCY PREPAREDNESS, BATON ROUGE, LOUISIANA

Mr. FONTENOT. Thank you. I would like to thank the Committee for inviting me today to speak on the events of the planning exercise known as Hurricane Pam as part of the Southeast Louisiana Catastrophic Planning Project. With this event, we began the process of trying to fully understand and prepare for the effects of a catastrophic hurricane hitting Southeast Louisiana. These remarks are a synopsis of the prepared testimony I have already submitted to the Committee.

In 1998, the Louisiana Office of Emergency Preparedness, now known as the Louisiana Office of Homeland Security and Emergency Preparedness, realized after Hurricane Georges that more planning was needed for post-landfall consequences after a major hurricane. A working group was convened consisting of Federal, State, and local participants to brainstorm the issues that Louisiana would be facing if a Category 3 or higher storm ever hit Southeast Louisiana. This work and a later meeting held in New Orleans in 1999 led to the development of a white paper, which outlined the planning proposal that was submitted to FEMA in August 2000 and then again in August 2001, asking for FEMA's help in planning and preparing for a catastrophic hurricane that could hit Southeast Louisiana.

In August 2001, FEMA headquarters awarded a contract to URS Corporation for catastrophic planning support. However, due to the events of September 11, 2001, there were many delays. In December 2001, a kickoff organizational meeting was held in New Orleans with FEMA headquarters, FEMA Region VI, and LOEP to organize this planning process. In January 2002, FEMA headquarters informed the State and Region VI that there would be no further funding for this project due to budget shortfalls. Following Hurricane Lili, the process was revived again for a short period in December 2002, but it also ended unsuccessfully.

In September 2003, there was a conference call with FEMA Region VI and FEMA headquarters to discuss the catastrophic planning. This led to a meeting on November 18, 2003, in New Orleans on this subject. Attending this meeting was a representative from the President's Homeland Security Advisory Council. At this meeting, LOHSEP and FEMA Region VI briefed the need for catastrophic planning, and he was astonished that as of that date, we had not completed this type of plan and promised to do what he could to help us get further funding for the planning process.

This brings us to the Southeast Louisiana Catastrophic Planning Project. On March 17, 2004, FEMA headquarters called FEMA Region VI and the State of Louisiana and informed us that there was funding for catastrophic planning. The very next day, LOHSEP and FEMA Region VI organized the Unified Command and steering committee. Later, a representative from FEMA headquarters was also added to the steering committee, as well. The concept was presented and approved by the Unified Command.

[1] The prepared statement of Mr. Fontenot appears in the Appendix on page 47.

On April 7, 2004, another meeting was held during the National Hurricane Conference in Orlando, Florida, to discuss the concepts with representatives from FEMA headquarters and to request that FEMA find a contractor to support this planning process.

On May 19, 2004, I was at FEMA Region VI working on the details of the proposed exercise. We were informed by FEMA headquarters that they intended to award the contract to IEM to support this planning project.

From the word "go," it was understood that this was not a typical exercise. In fact, when the concept was first given to me that we were going to have an exercise to develop a plan, I immediately disagreed. Usually, you write a plan and then have an exercise. However, when it was explained to me that we were going to take an exercise scenario which generated real consequences and real data and bring operational level people in so they could make decisions using the real data and consequences which could then drive the writing of a plan, I quickly got on board. I championed the fact that we were using operational people to write this plan because there are too many times a plan is written without taking the operational aspects into account and this leads to non-usable plans.

We initially met the contractor, IEM, at FEMA Region VI on May 20. At this meeting, we presented the exercise concept to IEM, and I pointed out, and FEMA Region VI agreed, that we had to work as a team and stick to our game plan to get this event accomplished in the time period available. We only had 53 days to put together something that would normally take 6 months to a year, and we couldn't push it back any further because August and September are the hot months for hurricanes in the Gulf.

We tried to involve local emergency managers as much as we could. For instance, when IEM developed a set of consequence estimates, the planning committee would meet with and poll local emergency managers to include them in the planning process from the beginning.

The Hurricane Pam exercise ran from July 16 to 23. On a typical day, the main exercise had six breakout rooms which had the same assigned topics for the entire week. Then we had three action rooms which were assigned topics on a day-to-day basis. Each day, the breakout rooms were responsible for writing a certain portion of the action plan based on the template that we had agreed upon with FEMA Region VI and LOHSEP prior to the event. The contractor had a facilitator and a recorder in every room to make sure that the room completed its task for the day. Also assigned to each room was a Federal and State lead who was responsible for briefing the Unified Command on a day-to-day basis. This process lasted for 5 days.

FEMA Region VI and LOHSEP expected that the action rooms would only produce the beginnings or the framework of a plan that would have to be fleshed out later. The breakout rooms were expected to produce more of a complete plan. However, we also knew that the breakout rooms would not develop a 100 percent answer.

Since this was not a standard exercise, there was no formal evaluation process. As I mentioned previously, this is an exercise designed to develop a plan, not test a plan. In my opinion, the exercise was very successful, not because it developed the perfect plans, but because it brought operational-level players to the table to begin the planning process. We never expected to come up with a 100 percent solution. It was always felt that if we had a 70 percent start, that we would be successful.

The scenario-based planning exercise, in my opinion, has produced the foundation of a very successful plan. However, due to the funding and time constraints, we had to be very selective about the topics covered during the main exercise and during the follow-on exercises.

The planning committee developed a scenario to show that it did not take a Category 4 or 5 hurricane to cause catastrophic damages in Southeast Louisiana. The National Weather Service Southern Region helped in the development of the weather scenario for Hurricane Pam. I wanted a slow-moving Category 3 hurricane that overtopped the levees of New Orleans, and the National Weather Service, working with the other NOAA partners, came up with the exact track and characteristics of the storm. The overtopping of the levees was included to cause the catastrophic flooding conditions from the storm surge.

The contractor was responsible for development of the consequences based on the storm scenario that the National Weather Service developed. All consequences were reviewed by the planning committee and the Unified Command. In addition, certain consequences were reviewed by the parish emergency management officials. This was to ensure the believability of the consequences and to get buy-in from the local emergency management officials. We knew that if the consequences weren't believable, then the focus of the players during the exercise would be on disputing the consequence numbers and not on developing the plans.

One of the primary things LOHSEP recognized at the conclusion of the exercise was that we needed to update our State Emergency Operations Plan to reflect the Federal Response Plan, now known as the National Response Plan. Essentially, we changed our State plan from a functional format to the Emergency Support Function, ESF, format, including the 15 ESFs associated with the National Response Plan. In this process of updating the plan, all function areas with the exception of one remained with the existing State agency that had been responsible for the function prior to the plan update, with the exception of the ESF–1 transportation, which was moved from the National Guard to the Louisiana Department of Transportation and Development. The final result of the plan update was that we had equivalent agencies at the State level talking to their Federal counterparts.

Initially, LOHSEP proposed to FEMA that we have a second major planning event like the Hurricane Pam exercise to focus on some of the areas that we did not get covered during the first exercise. It became clear after the first follow-on workshop that there would not be another large exercise due to funding. Therefore, FEMA Region VI and LOHSEP decided to use the second follow-on workshop to focus primarily on transportation, staging, and distribution of critical resources because it affected all the other plans in one way or another. Although I came to work for the Hurricane Pam exercise contractor, IEM, I recused myself from the Hurricane Pam follow-on activities due to my previous State responsibilities in line with counsel I received from the Louisiana State Ethics Board.

In spite of the funding, scheduling, and policy changes we faced with the Southeast Louisiana Catastrophic Planning Project, I felt that we had started down the right path. We still had a way to go, but we were heading in the right direction. It is my opinion that the scenario-based planning activities like Hurricane Pam are the way to go when trying to formulate plans to deal with catastrophic events. The realism that is brought to the table during these events really makes the planning feel more urgent.

I would like to thank the Committee once again for hearing my testimony.

Chairman COLLINS. Thank you. Mr. St. Amant.

TESTIMONY OF JESSE ST. AMANT,[1] DIRECTOR, OFFICE OF HOMELAND SECURITY AND EMERGENCY PREPAREDNESS, PLAQUEMINES PARISH, LOUISIANA

Mr. ST. AMANT. Good morning, Madam Chairman. Thank you for having me here today. I certainly appreciate the opportunity to speak before you and this group. Certainly, I would really like to thank you, as well, and your staff. They have really done a terrific job.

When I look at my notes and my statement I wanted to make, you two have covered it. I pray that someone is listening. The voices in the wind for too long have been out there. Time and again, we have expected and hoped that someone would hear our plea.

Mr. Fontenot just described to you something that I have here, a stack of documents going back, just these, from 1993 to current, and there are a lot more, telling of the horror stories that you have seen, and I really appreciate the fact that you have seen it because it is beyond description of what we are dealing with. So let me encourage you and the people in this great hall of justice that we are in, don't forget us. This is just the beginning. The horror story is not what has happened, it is what is still happening and continues to happen. I am going to get to that later.

Let me reassure and reaffirm some of the things that you have made known. Fair warning—Mother Nature has given us fair warning, and we have tried, as Sean said, to echo that. I remember telling, as the President of the Southeast Hurricane Task Force, stating this. If there is any significant loss of life, I would be the first to volunteer before any Congressional hearings, as I figured there would be some, because the fatality count could have been 100,000, not under 1,200. So for me, the Hurricane Pam exercise was really a success story because some of the lessons gleaned from that were some of the issues that we took back to our local jurisdictions to assist us in evacuation. Some of the things that we took back, we couldn't do alone, which is the reason I stated I would be glad to testify before anybody because we needed the continuing support to have a Pam exercise, and my group of directors representing Southeast Louisiana and some 15 jurisdictions had to

[1] The prepared statement of Mr. St. Amant appears in the Appendix on page 56.

beat on the desk a little bit to make sure that it would be funded because someone didn't think that it was important.

Well, in any case, we got it done, but Mother Nature has a sick sense of humor. She showed to us that I will hit you before you are ready. I hear this diatribe about 50-foot levees or what they call Category 5 levees that are being planned or being cried for and being asked for. My experience in emergency management tells me this. You build a 20-foot levee, Mother Nature will give you a 25-foot storm surge. The maximum envelopes of water, the loss of the wetlands, we can blame everything and his brother for what has happened, but the fact of the matter is, due to the soil subsidence, due to the loss of our wetlands, we knew in this business that this was coming. We tried to say the words, this is coming, time and again.

One of the documents I wanted to show you today was this one, dated 1994. It gives you the exact scenario of the worst case scenario that could happen. It was never a case of if, it was a case of when. This document from the Government Accounting Office tells you what is going to happen.

But the fact of the matter is that due to the Pam exercise, we really got a little bit better about getting some people out of harm's way. I would hate to think what would have happened had it not been. Maybe the fatality counts, as I said, would have been greater.

Dr. Bob Sheets, former Director of the National Hurricane Center, gave this warning. I also happen to have this on video. New Orleans is the worst case scenario in the continental United States, surrounded by water, at or below sea level, 1.6 million people, with lack of infrastructure to evacuate in a timely manner. I submit to you that is not my only concern for hurricanes, something that we may have 2 or 3 days to see and to prepare for and respond to or evacuate from. My concern is what happens if we have some other type of event that doesn't have that much notice, maybe a chemical spill that we may have to evacuate people in the short term.

These are the considerations, and let me say one other thing further. Let us suppose Miami, Houston, Washington, DC. We are talking about the Hurricane Pam exercise that was supposed to raise the awareness level of a major catastrophic event happening in any major city, not just New Orleans.

Folks, we were lucky. There are some things that I am going to recommend, some of which is, if it is not broken, don't fix it. My fellow directors from the State of Louisiana have always said, FEMA used to be a good organization, but somebody decided we were going to put it under Homeland Security, for whatever reason. Personally, I feel that the Federal Coordinating Officer, and the Defense Coordinating Officer, the State Coordinating Officer, working together, can resolve most of the problems. I remember some of our response that we did for the Andrew situation—I had been there a few years—and I thought they were very good because you had the right people, the communications, and the coordination.

But the fact of the matter is, when you build top-down approach, you have got people in cubicles at the top giving directives to very few at the bottom. Nothing happens. If you get people who don't listen to the warnings that we are trying to say and they go unheeded, then nothing happens.

So I believe in us being more proactive than reactive. I believe in us doing what needs to be done, and if you want to find out, ask the people who were involved. I think this is why I admire your tenacity in searching and seeking the people who should know what this is all about.

Let me assure you of one thing. We will recover and this will happen again. Will we be prepared? I submit we need to lead, follow, or get out of the way, and I submit that for a simple reason. The loss of life this time was just a wake-up call. As sad and tragic as it is, this was not a direct hit. This was a glancing blow. If the eye wall of the hurricane had been 12 miles further west, I would not be here giving this testimony and a lot of other people wouldn't be on the face of the earth. The fact of the matter is, this glancing blow did cause some overtopping and some levee failures, but the City of New Orleans would look like the lower part of Plaquemines Parish, where nothing would exist, had it crossed over the City of New Orleans.

The gallant response, the efforts that were made were hindered by the lack of communication, the lack of coordination and damage assessment. No one ever anticipated that 100 percent of the communications that we had—if you can't communicate what your problems are, if you can't reach out, then your response is hindered. I had people 3 weeks after the storm that were amazed to find out that I was still alive because I couldn't communicate. When we finally had satellite phones delivered to us by the State, they were of no use because they wouldn't work.

I think the President has acknowledged that communications interoperability has got to be the most important essence of our response and recovery and preparation. I happen to agree with that. When you don't have anything, and two tin cans and a string doesn't cut it and carrier pigeons, as the former director used to say, don't want to fly in bad weather. It is tough to cry, I need help, when no one hears you.

Again, I want to wish you luck and continuance on your endeavor to try to reach a conclusion, and I just hope someone will listen to what is being said here. This is an opportunity to go forward and to make sure that the next time it happens, as it will, we will be better prepared. I thank you for this opportunity.

Chairman COLLINS. Thank you. Ms. Beriwal.

TESTIMONY OF MADHU BERIWAL,[1] PRESIDENT AND CHIEF EXECUTIVE OFFICER, INNOVATIVE EMERGENCY MANAGEMENT, INC., BATON ROUGE, LOUISIANA

Ms. BERIWAL. Madam Chairman, Senator Lieberman, Members of the Committee, thank you very much for the opportunity to testify in front of you on catastrophic planning for Southeast Louisiana, called Hurricane Pam.

A lot of the points that I was going to make, the previous witnesses have already mentioned those, so I won't belabor you with those issues. I want to start by making, first of all, a distinction between what Hurricane Pam was and was not. There has been a

[1] The prepared statement of Ms. Beriwal appears in the Appendix on page 58.

lot of confusion and chaos about what Hurricane Pam was and wasn't.

First of all, if you go back to traditional emergency planning, in traditional emergency management, you have essentially a quality cycle that starts with planning. You prepare a plan. Those plans are generally prepared by either one person or a small committee of five or six people. It takes 6 to 12 months to prepare a plan. And then you take that training, and all of the people with primary responsibility for execution of the plan go through a training cycle. That might take another 12 months or so to do. And then you have an exercise. Planning for the exercise generally takes 6 to 12 months to execute. You have an exercise for a few days, and then your report might come out as soon as 2 or 3 months after the exercise or sometimes as long as a year after the exercise. This whole quality cycle takes somewhere between 2½ and 4½ years, depending on the complexity of the topic and the complexity of the region that is involved.

This is not what Hurricane Pam was. Even though Pam was called an exercise, it was not a traditional exercise in the sense that there was a plan in place and that we were going to exercise the plan. I don't mean to imply that there were no plans in place. The 13 Southeast Louisiana parishes that participated in Hurricane Pam all had emergency operations plans. Several of them had hurricane plans. The State of Louisiana had plans. There were 20 State agencies involved. Many of them had emergency operations plans in place. And, of course, the National Government had the Federal Response Plan when we started and the National Response Plan further on into the process. So everybody had legally constituted plans.

The effort for Hurricane Pam was to create a bridging document between all of these local plans, the State plans, and the National Response Plan. This is a term that was used widely during Hurricane Pam in the many workshops we conducted, is to create a bridging document that will be addressing just catastrophic events.

Most plans deal with a gamut of hazards, everything from chemical spills, radiological events, hurricanes, floods, and tornadoes. The intent of Hurricane Pam was to create a plan for a catastrophic event, a specific event. As some of the previous witnesses have testified, the intent was to create a sense of reality. When we were working with this project, we were trying to describe a worst case but plausible event. That is the slogan that we had. It has to be plausible because it was very important that the exercise not degenerate into questioning the data on the basis of which of the plans would be developed.

We started on May 24, 2004, when we were awarded the contract. Actually, we had verbal notice to proceed from FEMA earlier than that, and we met with the FEMA Region VI and LOHSEP in Denton, Texas, to plan out this exercise. We had 53 days to put an event together of considerable complexity and magnitude. We understood that, but we were dedicated to making the Hurricane Pam workshop a success.

It was an 8-day exercise, and in the 53 days prior to the event, we cascaded from the slow-moving Category 3 storm that you, Madam Chairman, mentioned and that was briefed by the National

Weather Service. We took the data from the National Weather Service and their slosh model and predicted a series of consequences. I would like to tell you some of these consequences and compare them to Katrina. I know that there has been a lot of discussion about how similar these consequences were.

We added 20 inches of rain into Hurricane Pam prior to the event to create catastrophic conditions both from storm surge and from rain. As you might know, nine of 10 deaths that occur in hurricanes are due to storm surge and due to drowning from rain and storm surge. So we wanted to create 10 to 20 feet of water in the City of New Orleans, which would constitute a catastrophic scenario for Southeast Louisiana.

We overtopped the levees. We did not breach them. We also looked at the Louisiana offshore oil port, and as you know, Senators, the significance of that oil port is that it handles 12 percent of the crude oil of the United States, and that LOOP port would close prior to the storm and would come back 2 or 3 days after the storm.

To give you comparable data for these consequences, in Hurricane Katrina, there was actually 18 inches of rain. The levees were overtopped as well as breached in places. Louisiana Offshore Oil Port (LOOP) did close for 5 days before and after the storm.

We predicted that nine refineries would shut down during the storm. Actually, seven refineries shut down. We predicted that 57 chemical plants would be flooded and shut down. Over 50 plants were flooded and shut down.

We predicted that 1.1 million people would be made homeless from the storm. The actual number is about 1 million.

We expected that Leeville Bridge on Louisiana Highway 1 to the west of the city would collapse, since we had the track of the hurricane on the west of the city putting the northeast quadrant, which is the most damaging part of the storm, directly over the City of New Orleans. In fact, the New Orleans Twin Span bridge collapsed to the east of the city since the storm track of Katrina was to the east.

We expected that 786,359 people would lose electricity at the initial impact; 881,400 people actually lost electricity after impact.

We predicted that there would be 12.5 million tons of debris that would be generated. The estimates right now are that there are 22 million tons of debris, 12 million tons just in the City of New Orleans itself.

We predicted that there would be extensive coastal marsh erosion. The initial indications are that Louisiana lost a year's worth of coastal marsh erosion in the one day of Katrina's impact. Just so that you understand what the significance of that is, in the 33 minutes since the start of this briefing, an area the size of the greater Washington, DC, area disappeared in Louisiana, and it is continuing to disappear at the rate of 25 square miles a year.

We also said the sewage treatment facilities would not work in the metropolitan area, which is exactly what happened in Katrina since they are powered and the power would be lost.

We expected that 233,986 buildings would collapse and 250,000 homes are considered to be destroyed from Katrina.

We expected that 15 percent of the 13 parish hospital supply would be affected and some of it would be completely destroyed. At present, there is no medical system available in the City of New Orleans for those that are not insured.

We expected that there would be $40 billion in damages to commercial and residential structures in Louisiana, and the Insurance Institute has estimated that the damage to commercial and residential structures is between $20 and $65 billion.

We expected that there would be 61,290 deaths. Fortunately, we were wide off the mark on that one. At present, we have 1,100 people known to have died in Louisiana. Another 3,000 to 4,000 are still missing and not presumed dead as yet.

I would like to move away from the consequences, but just in closing on that particular topic mention that developing these consequences was very important. We wanted to create a sense of urgency. We wanted to create a sense of realism in the exercise which generally does not inform a planning process when you are dealing with emergency planning. Because we are all mortal beings, we don't like to look at the face of death and disaster, and most planning tends to look at the event that you can manage, not the events that you can't manage. The Hurricane Pam exercise was designed with detailed consequences down to the parish level for each of these data elements. We actually had data on how many people would be affected by parish so that each of the individual parishes and the State and FEMA would have tactile information at their fingertips that they could use in planning.

How much of that got used? I know there has been a lot of confusion on this topic, too. It seems from some of the reports that Hurricane Pam did not have any effect. I would beg to differ. We did have a lot of effect, and I will further on talk a little bit about what I think we could have done better.

Talking about the things that got used in Katrina, first of all is the response rate. In Hurricane Pam, we projected that 36 percent of the 1.9 million people, that is 1.7 million residents of Louisiana and 200,000 tourists, would actually evacuate. That is 36 percent of 1.9 million people. That would have left a considerable number of people in the 13-parish area. Why did we project such a low number? Because history has indicated from Hurricane Georges in 1998, Hurricane Ivan in 2004, as well as prior storms, that not enough people leave in the face of a storm.

In Hurricane Katrina, now I am going to give you information that is not scientifically validated as yet, but at least indications are that 80 to 90 percent of the people in the 13-parish area left that region. That is equivalent to ringing the bell in emergency management in terms of evacuation for a large metropolitan region. It has not occurred before. The most validated information on prior storms where there has been a high rate of evacuation was Hurricane Hugo, where 81 percent of the people evacuated in the face of that storm.

I think most of the credit for that goes to the National Weather Service, Dr. Mayfield at the National Hurricane Center, and the media for publicizing the impending storm. But I think that we can take a small measure of comfort in the fact that some of the actions of the State, Federal, and local officials were motivated by the high casualty count of Hurricane Pam and the consequences projected in this particular planning exercise.

We also developed a search and rescue process called the lily-pad operation where people would essentially be plucked from the flooded areas, brought to the land-water interface, and from there they would be taken by another set of people to the shelters or to medical facilities where they would be treated or taken care of. And when I say "we," I mean the participants of Hurricane Pam and IEM. We did not see a division between the company and the customers that we serve. So this was brought up by the participants. They developed this concept. You saw that on CNN and FOX News during Hurricane Katrina operating to save lives.

In the data that they are gathering right now post-Hurricane Katrina to compare Hurricane Pam and Hurricane Katrina consequences, we think that somewhere between 60,000 and 100,000 people went through the search and rescue method where they were rescued from rooftops and from flooded buildings and brought using the lily-pad method.

Another issue that I wanted to mention is the TMOSA, which is the Temporary Medical Operations Staging Area. Those were effectively used in Katrina. We had predicted that there would be three needed. Three of them were operational, two real ones at LSU and Nicholls and then the other one was actually the New Orleans Airport, which effectively became a TMOSA.

Let me close quickly and mention to you a few things that I think need to be done better. I have worked in emergency management and homeland security for 26 years now, and I think that we really need to look very carefully at how we do emergency management and homeland security. We are spending about $1 million a minute in homeland security and emergency management in this country. I think we need to demand better results.

The first thing that I would mention is that we need to have an outcome-based emergency management homeland security process, something where elected officials can say, this is what I expect the outcome to be, and then emergency management and homeland security are tasked with delivering those outcomes. This is no different from the Government Performance and Results Act or the President's Management Agenda, which has been mentioned in the last several administrations. We need to apply the lessons of that management philosophy to emergency management.

Second, I think that we need in emergency management a way to measure protection. We came up with a lot of innovations in Hurricane Pam. A number of them were used at a non-scientific count, but about 75 percent of those things got used in Katrina, yet the results were deemed unacceptable by the President, by the media, and by the American public. We need a way to calculate protection. I would not want to run my company without knowing what the profit and loss statement was. How can we run emergency management without knowing what level of protection we are providing?

The third thing, we need a reliable and mature emergency management process, one that creates a professional discipline out of this field.

And fourth, we need a way to do sustainable development in our community so that we do not have problems like coastal erosion and other such factors that affect the vulnerability of the region to natural as well as unnatural disasters.

Thank you very much.

Chairman COLLINS. Thank you.

Mr. Fairley, I would like to start my questioning with you. Committee investigators were told by a former colleague of yours, Mark Wallace, who also worked on designing Hurricane Pam, that the effectiveness of the project was greatly diminished by the poor attendance of key decisionmakers, and I would contrast that to an exercise that Senator Lieberman and I observed last year called the TOPOFF exercise where the Governors of New Jersey and Connecticut and the Secretary of Homeland Security were directly involved in a scenario simulating a terrorist attack.

Mr. Wallace told the Committee that had the Director of FEMA, the Governor of Louisiana, and the Mayor of New Orleans participated in the scenario, that binding agreements could have been reached, that there would have been a better understanding of the responsibilities of the entities and the plans they were to follow. Were efforts made to bring the high-level key decisionmakers like the FEMA Director, the Governor, and the Mayor into this process?

Mr. FAIRLEY. I can only speak from the FEMA perspective. I will then let my colleagues answer for the State and local. When we put the, what we call the leadership committee or leadership group together, we extended an invitation for FEMA headquarters involvement. Naturally, we requested the highest level that we could get. I am not aware of what decisions were made as to who would attend. We did receive people from headquarters who were in lines that could make decisions and could make recommendations.

I would never argue with anyone, the higher the person you have at your meeting, the less meetings you would probably need to have or the more decisions you could have made on the spot, but we felt comfortable with the leadership that came. I think, naturally, you would always like to have more, but we felt comfortable that the people there could relay back what they found, what they saw, and assist us in getting decisions made.

Chairman COLLINS. Mr. Fontenot, in your testimony, you discussed talking about the need to improve planning in Louisiana way back in 1998. And during the next nearly 6 years, until the Hurricane Pam exercise was actually funded and took place, there was a lot of communication back and forth between the State and FEMA, and I have looked at the documents which present a very compelling case for the need for this kind of catastrophic planning.

Could you give the Committee a better understanding of what happened during that 6-year period? Why did it take so long from when you first identified a very real and compelling need and the time that the exercise was actually held?

Mr. FONTENOT. First, I think that it is important to recognize and important to state that we weren't just sitting on our hands in that 6 years. I mean, we were actually doing planning on our own and with the local governments and with our State counterparts, trying to figure out some of the aspects that we knew that we could handle on our own.

Second, I think that in my written testimony, and in my verbal testimony earlier, I talked about the different conference calls we had and the different meetings we had and some of the conversations that we had with FEMA, FEMA Region VI, and FEMA headquarters. It also needs to be pointed out that FEMA Region VI was always a very willing participant in anytime we went to FEMA headquarters to request funding for these exercises to take place and this planning event to take place.

I know that on several occasions, my former Assistant Director of Emergency Preparedness came to Washington and met with Director James Lee Witt and then Mr. Allbaugh, requesting that we do this planning event. I know that the gentleman that had the position of Chief of Plans before I did, or Chief of Planning, Training, and Exercises, before I took over for him, I know that he spoke to Mr. Allbaugh about this planning cycle and the importance of it.

I know that a lot of things that went on, I wasn't always at every meeting that was conducted and discussions of this, so I really can't answer what took so long other than we pushed as much as we could and pushed as much as we thought that we had the capital to push without really upsetting people for pushing too hard.

Chairman COLLINS. Let me talk to you about the question that I asked Mr. Fairley. Do you think Hurricane Pam's effectiveness would have been improved if the Director of FEMA, the Governor of Louisiana, and the Mayor of New Orleans had directly participated in some of the simulation?

Mr. FONTENOT. Well, I think Wayne answered quite adequately about the Director of FEMA, so I will concentrate on the Mayor of New Orleans and the Governor. I can tell you that there was at least one briefing to the Governor's office prior to Hurricane Pam where the Director of Emergency Management for the State of Louisiana, which is the Adjutant General, was briefed and then he briefed the Governor's office. Unfortunately, there was a couple layers of management above me, so I really never—I never had any direct dealings with the Governor or her office, so I don't know exactly what the conversations were and what happened between them, the Adjutant General, and the Governor's office as far as inviting the Governor or her staff to the exercise. They were more than welcome to come, but I don't know what those dealings were. You would have to ask him.

As far as the Mayor of New Orleans is concerned, we did invite the parish emergency managers from all 13 parishes in the Southeast Louisiana Task Force, and we invited them to bring whoever they wanted to bring with them. It was up to the emergency manager on who they chose to bring with them and who they chose not to bring with them. However, there was financial constraints, as well, and we had over 300 participants at this exercise. It needs to be pointed out that we were pretty much pushing the envelope of how many people we could have handled without going to an off-site place to hold the exercise, which we could have done, it just would have cost us some more money. So that is my answer.

Chairman COLLINS. Ms. Beriwal, you testified that not enough people evacuate prior to a storm hitting an area, despite public officials urging it, despite even mandatory evacuations. Your scenario predicted that hundreds of thousands of individuals would not evacuate. Did it concern you, then, that pre-storm evacuation was excluded from the Hurricane Pam exercise, given that, as you said today, not everybody or not as many people as should evacuate do so?

Ms. BERIWAL. Pre-storm evacuation is actually a big problem. In a nutshell, the issue is that about 24 hours prior to landfall, there is a 50–50 chance that the storm is actually going to strike the region to which it is destined, and people make their own determinations. I would like to say that under carefully controlled circumstances, people do damn well as they please, and so each individual family and each individual person in an area decides whether they are going to evacuate or not evacuate.

However, about 50 to 60 years of emergency management literature tends to indicate that people leave if they are told by credible local officials to leave. Since I have worked off and on with the City of New Orleans since the 1980s, they have never ordered a mandatory evacuation for the City of New Orleans until Hurricane Katrina. So you cannot have a high percentage of people leave unless you have a mandatory evacuation ordered by people that others recognize and who essentially stand up and say, "I am the mayor or the parish president, and I order a mandatory evacuation of this area."

Chairman COLLINS. But if you could answer my question more directly, did you express concern when the State decided to exclude pre-storm evacuation from the exercise?

Ms. BERIWAL. No, I did not because for the longest time, at least in my knowledge, for the last 20 to 25 years, every exercise for hurricanes in Southeast Louisiana has focused only on the evacuation question, and Hurricane Pam was expected to be the first post-storm exercise to look at response post-storm. So the fact that 3 days of the 8-day event were devoted to pre-landfall and 5 days to post-storm seemed like an appropriate thing to do.

Chairman COLLINS. Thank you. Senator Lieberman.

Senator LIEBERMAN. Thanks, Madam Chairman. Thanks to the four of you. Your testimony has been very helpful, very troubling, I guess, insofar as you were all involved in an effort to get ready for what came in Katrina, and to some extent, I hear you saying it helped, but to a lot of other extent, it didn't put all those in government in a position to diminish even further the consequences of what happened.

I think I will pick up, just so I understand exactly, where Senator Collins left off, which was this fact that—I will start with you, Mr. Fairley—while FEMA agreed to propose this exercise Pam in 2001, it doesn't get underway until 2004. In that time period, can you tell us, to the best of your knowledge, the reason for the delay?

Mr. FAIRLEY. Senator, the only logical reason I can give you is that there were not funds available.

Senator LIEBERMAN. Yes.

Mr. FAIRLEY. Not being part of the budgeting process in Washington, I am not familiar with all the little nuances. However, I know that if we propose a project and it is late in the fiscal year, the agency's budget generally has already been set and has worked its way through all the committees. So this request for project and project funding goes into the next budget cycle. So that could account for up to 2 years sometimes.

Senator LIEBERMAN. Mr. Fontenot, in your opening statement, you mentioned, and the question, I just want to get it clear, that a White House representative attended a meeting in New Orleans in November 2003 and "was astonished that as of that date, we had not completed this type of plan, and promised to do what he could to help us get funding for this planning process." I didn't get that clear, was that Joe Allbaugh or was that somebody else, or do you not recall?

Mr. FONTENOT. No. I was at the meeting. Actually, it was Retired General John Gordon, and he was on the advisory council, Homeland Security Advisory Council.

Senator LIEBERMAN. OK. So——

Mr. FONTENOT. He was the gentleman that we were briefing.

Senator LIEBERMAN. Right. And you have some reason to believe that he went back and helped to facilitate the funding that resulted in Pam?

Mr. FONTENOT. That was my understanding, yes.

Senator LIEBERMAN. OK. Let me now go to the question of pre-storm evacuation. As Senator Collins said, as we have seen it in the records, the Committee has obtained early proposals that became Pam sought funding to study the problems of pre-storm evacuation. Later, this was removed from the planning exercise. Mr. Fairley, do you know why that happened?

Mr. FAIRLEY. Yes, sir, in general. When we came together in the various meetings to discuss items that would result in planning topics, one of the things that we all experienced was for every question we asked, instead of coming up with an answer, we came up with five more questions.

Senator LIEBERMAN. Right.

Mr. FAIRLEY. And we realized very quickly that if you look at putting a catastrophic plan together starting with pre-landfall, then response, and into recovery, that we could not finish it in one session. We may not be able to finish it in one year or several years. So we met with the State and said that we needed to look at something that we could handle in a short time frame or shorter time frame, and it was decided among everyone that response to the hurricane would probably be more appropriate than to worry about long-term recovery issues, which the response would probably dictate.

Senator LIEBERMAN. Yes.

Mr. FAIRLEY. So we went into the phase of eliminating items that were not considered response.

Senator LIEBERMAN. Looking back, and I know hindsight is always clearer than foresight, do you wish that you had included in Pam some element regarding pre-storm evacuation, which was obviously a big problem in Katrina?

Mr. FAIRLEY. Yes, sir. Hindsight says that evacuation was a very important element. We went on the basis that local and State law requires local and State evacuations, and we would support that.

Senator LIEBERMAN. Understood. Mr. St. Amant, can you tell us whether the Federal agencies in the Hurricane Pam exercises were advised that the City of New Orleans and surrounding areas had no effective way to evacuate people without personal transportation or were lead agencies advised in Pam of the city's efforts to prepare long-term for pre-storm evacuation?

Mr. ST. AMANT. Absolutely, sir.

Senator LIEBERMAN. They were? What was your understanding, if you had one at that point, of any possible Federal role in pre-storm evacuation of a catastrophic hurricane?

Mr. ST. AMANT. There is no question that no area with 1.6 million people, with the lack of intermodal infrastructure, can move in a very fast or efficient manner on its own——

Senator LIEBERMAN. Right.

Mr. ST. AMANT [continuing]. Specifically, not any poor States and poor areas such as Southeast Louisiana. To give you an idea, in my jurisdiction, I am divided by the Mississippi River. I have one road on each side to get out and get in. That is it. I have to get through two other jurisdictions to get my people safely out of the risk area. The bottom line, at the end of the day, there is no way that New Orleans, Jefferson, or anybody in that region is going to be able to meet this challenge on its own.

We made specific knowledge known to them that as of the 1990 census, the numbers of people in the region who are dependent upon regional transportation because they don't have their own automobiles. This effort and the surrounding challenges of the lack of intermodal transportation resources caused me great concern, not because of the hurricane that may give you 2 or 3 days to move, but short-term notice of evacuation, regarding the resources necessary, sir.

So I will tell you this. Yes, I was there, and by the way, yes, my parish president did attend some of these sessions. He didn't have to be there. That is what he hires me to do, to advise him, to make sure. I answer directly to one man, not a committee, and that is why we tried to practice what we preach.

Senator LIEBERMAN. Mr. Fairley, from the FEMA point of view and insofar as you know from the Department of Homeland Security point of view generally, what, if any, changes occurred in response to the Hurricane Pam exercise, including the sense that we get to some extent—Mr. St. Amant just testified to it—it certainly comes through the Pam report and plan that in the event of a catastrophic hurricane, State and local first responders were going to be overwhelmed? Were there any changes in Region VI, the one that covered New Orleans, in terms of FEMA preparedness or plans to respond?

Mr. FAIRLEY. Yes, sir. In the frame of mind, there was great changes. I think a lot of us as a result of these sessions walked away seeing holes and gaps and fearing that we would not have things ready in time. As Jesse said, it was not a matter of if but a matter of when. I think we all were hoping that we could buy one more year.

Senator LIEBERMAN. Yes.

Mr. FAIRLEY. Yes, sir, we did try to speed things up. We tried to change directions. Not everything, of course, was available. At the time of the very first session, we were not sure that there would be a second session, so we were looking at trying to get things done on, I hate to say a fast pace, but a faster pace than normal. As it turned out, we did get a few extra sessions. So, yes, there were some changes of philosophy. When we worked with the locals and some of the State agencies, we realized that what we had always thought to be standard practices or were plans that were solid were, in fact, there were gaps in them, and it was through that cooperation that we discovered that.

Senator LIEBERMAN. With respect to you, and this is really a question to be asked of those higher up in FEMA who we will have before us, in response to the Hurricane Pam exercise, which was, as I said earlier, actually eerily predictive, what was necessary was more than a change of frame of mind. In other words, ideally, there would have been more action put into effect. I guess the ultimate question is why was FEMA and the rest of the Federal Government so slow, certainly appearing to me, in responding to both the clear oncoming of Katrina and then in responding once it hit?

You know what, I don't even need to have you answer that question. That is what I am going to ask. But the point is, on the record, we don't see enough of a response certainly in the days before and immediately after Katrina hit landfall to exactly the lessons of the Hurricane Pam exercise and the plan itself, a very impressive, extensive, and detailed document.

I want to come back on my second round and ask some more about what happened to the plan. Thanks, Madam Chairman.

Chairman COLLINS. Thank you. Senator Levin.

OPENING STATEMENT OF SENATOR LEVIN

Senator LEVIN. Thank you, Madam Chairman, and to you and Senator Lieberman, all of our thanks for your extraordinary and typical tenacity in digging into this issue and all of its ramifications. The Nation is again truly in both of your debt for what you are doing here, and hopefully, it is going to help us deal with future catastrophic situations.

I am a little uncertain on the question of what happened immediately prior to Katrina. Given the previous level of planning, given the previous studies that have been done, is it clear who was responsible primarily for the evacuation both pre-Katrina and post-Katrina, that rested in State and local governments rather than FEMA in terms of primary responsibility for evacuation? Mr. Fairley.

Mr. FAIRLEY. Senator, in my mind, yes, it was. The State of Louisiana law requires that Louisiana and its parishes prepare for evacuation from events. Our role is to support that as requested or as directed.

Senator LEVIN. And Mr. Fontenot, is that your understanding?

Mr. FONTENOT. Yes, Senator, that is my opinion. I would also add that I think that New Orleans also knew that was the case because before I left State Government there was a meeting to discuss how the State could help them with that role, and——

Senator LEVIN. OK. And Mr. St. Amant, is that your understanding, as well?

Mr. ST. AMANT. The Louisiana Disaster Act clearly delineates the responsibilities of the emergency managers and parish presidents, etc. Yes, sir.

Senator LEVIN. Thank you. Now, Exhibit H,[1] you all have exhibit books, it is called the "New Orleans Hurricane Shelter," and on the first page it says, "Even under the best conditions, evacuation will leave at least 150,000 people in harm's way." I think it is the fourth page has something which is a document headed, "Louisiana Superdome: Refuge of Last Resort." Do you see those documents, or that document, Exhibit H? Whose document is that? Is that a FEMA document or a parish document, a State document, what is that, does anyone know?

Mr. FONTENOT. If I may, I think this is a New Orleans Parish document. It is not a State document, and I am pretty sure it is not a Federal document, but I will let Wayne talk to that.

Mr. FAIRLEY. No, sir, it is not a FEMA document.

Senator LEVIN. It states here that not all citizens may be able to evacuate due to medical infirmity or dependency. It makes the statement that more than 57,000 households in New Orleans do not have access to an automobile and have not made adequate arrangements for evacuation. This is a life and death situation, that the Superdome may be mobilized as a refuge of last resort. So that was clearly known to whoever prepared that document, and I think in general is it fair to say that it was anticipated that a catastrophe of this scope could occur in New Orleans? Is that a fair statement, that all of you agree that it was anticipated that a catastrophe or a hurricane of this size and this impact could and probably or perhaps would occur in New Orleans? Is that a fair statement?

Mr. ST. AMANT. Absolutely. It was our worst case nightmare.

Senator LEVIN. All right, but I want to just go quickly along. Mr. Fairley, is that a fair statement?

Mr. FAIRLEY. Yes, sir.

Senator LEVIN. And Mr. Fontenot.

Mr. FONTENOT. Yes, sir. We always said it was not a matter of if, but when.

Senator LEVIN. OK. Ms. Beriwal.

Ms. BERIWAL. Yes, sir.

Senator LEVIN. Now, after the catastrophe, we have Secretary Chertoff saying that this catastrophe exceeded the foresight of the planners and maybe anybody's foresight. How could he make that statement? Mr. Fontenot, I will start with you.

Mr. FONTENOT. I have no idea. You will have to ask Mr. Chertoff why he made that statement. I think that we have shown for years, we have been yelling about this potential disaster.

Senator LEVIN. Mr. Fairley, have you talked to Mr. Chertoff about this, or——

Mr. FAIRLEY. No, sir, I have not.

Senator LEVIN. FEMA is in his Department, as I understand it, or still there. Mr. St. Amant.

Mr. ST. AMANT. July 22, 2005, quoted in the *Associated Press* where I stated when they released the study on the evacuation, behavioral study by UNO-New Orleans, that the results would be beyond comprehension. Anybody who has seen this, as most of you have, know what I am talking about. It is beyond comprehension.

---

[1] Exhibit H appears in the Appendix on page 100.

Senator LEVIN. It also was anticipated, was it not?

Mr. ST. AMANT. Absolutely, but it still wasn't the worst case. If you think this is bad, no, it is not.

Senator LEVIN. But nonetheless, a catastrophe of this scope at least was anticipated.

Mr. ST. AMANT. Should have been.

Senator LEVIN. Ms. Beriwal.

Ms. BERIWAL. Senator, I cannot comment on what Mr. Chertoff may or may not have known. I have no knowledge of it.

Senator LEVIN. But from your perspective, a catastrophe of this scope was clearly anticipated, was it not?

Ms. BERIWAL. Yes, sir.

Senator LEVIN. Now, immediately prior to the storm, on Saturday, if you look at Exhibit F,[1] I guess this goes to you, Mr. Fairley. The FEMA staff at headquarters gave a briefing using a five-page Power Point, which is Exhibit F, and what that exhibit said, and this is the Saturday prior to landfall, that the Pam exercise projection is exceeded by Hurricane Katrina real-life impacts. Storm surge could greatly overtop levees and protective systems. Potential fatalities, 60,000. Incredible search and rescue needs of over 60,000 persons. Displacement of a million-plus population. Do you know, Mr. Fairley, who gave this briefing?

Mr. FAIRLEY. No, sir, unfortunately, I do not. On Saturday, August 27, at around 12 noon, I was packing a suitcase, trying to get a plane to Baton Rouge, Louisiana, so I was not privy to this.

Senator LEVIN. Thank you. Fair enough. So you wouldn't know who was briefed?

Mr. FAIRLEY. No, sir. I would assume that this was a briefing in the FEMA NRCC.

Senator LEVIN. OK.

Mr. FAIRLEY. Excuse me, I'm sorry about using the initials, the NRCC, which is the National Response Coordination Center. But I'm not sure which official was making it.

Senator LEVIN. Do any of you know who gave the briefing and who was briefed?

Mr. FONTENOT. No, sir. I have no clue.

Ms. BERIWAL. No, sir.

Senator LEVIN. All right. Now, the next day, or the same day, Exhibit K,[2] there was a computer simulation run at the National Infrastructure Simulation and Analysis Center. Is that NISAC, does that sound correct?

Ms. BERIWAL. NISAC.

Senator LEVIN. OK, at NISAC. This is part of the Department of Homeland Security, and it was performed on August 27, this computer simulation, and it is Exhibit K, and there was an update performed on August 28, and this was delivered to the White House Situation Room at 1:47 on Monday morning, August 29. This NISAC report stated that the potential for severe storm surge to overwhelm Lake Pontchartrain levees is the greatest concern for New Orleans according to the NISAC report. So Homeland Security knew prior to the breach of the levees, at least a number of hours

[1] Exhibit F appears in the Appendix on page 95.
[2] Exhibit K appears in the Appendix on page 104.

before the breach of the levees, that this was the greatest concern for New Orleans. Do you know where the NISAC folks got that terminology, Mr. Fairley?

Mr. FAIRLEY. No, sir, I do not.

Senator LEVIN. But is it fair to say that in terms of the impact of a severe, catastrophic storm that it was known that the breach of the levees could be one of the impacts?

Mr. FAIRLEY. Yes, sir, I think in all of the planning scenarios and past disasters that we always knew that a breaching or an overtopping of the levee could lead to——

Senator LEVIN. Either one?

Mr. FAIRLEY. Either one.

Senator LEVIN. And do you know who got that report at the White House?

Mr. FAIRLEY. No, sir, I do not.

Senator LEVIN. All right. Does anyone here know?

Mr. ST. AMANT. No, but I wish they would have shared it with us. That might have been nice.

Senator LEVIN. Are you familiar with this?

Mr. ST. AMANT. Not at all.

Senator LEVIN. Is anyone familiar with Exhibit K?

Ms. BERIWAL. No, sir.

Senator LEVIN. Mr. Fontenot.

Mr. FONTENOT. No, sir.

Senator LEVIN. If you look at page 37 of that exhibit—by the way, we also have the statement of the President that he says he doesn't think anyone anticipated the breach of the levees. Now, I don't know how he can say that given the fact that everybody anticipated the breach of the levees according to you folks, but I guess that is something the White House is going to have to respond to. He said that on Thursday, September 1, on *Good Morning America*. "I don't think anyone anticipated the breach of the levees," when it is obvious that everybody anticipated that was a realistic possibility. But go to page 37.

Mr. FONTENOT. Sir, we don't have a page 37.

Senator LEVIN. All right. Do they have this exhibit? I am out of time anyway. I will have to get to that in my second round. Thank you.

Chairman COLLINS. Senator Carper.

OPENING STATEMENT OF SENATOR CARPER

Senator CARPER. Thanks, Madam Chairman. Our thanks to each of our witnesses. We are delighted that you are here. We appreciate very much your testimony and putting some light on these issues that we wrestle with.

I think I would like to start off with a question for each of you, if I may, and then I have a couple of individual questions. We learned a number of lessons about the gaps in planning during the Hurricane Pam exercise, but there are still quite a few questions that we know we need to follow up on. I guess my question for all of you would be this. If you had to do it all over again, how would you ensure that the lessons learned during this exercise were better translated into your particular agency or entity's emergency plan?

Mr. FAIRLEY. That is a very good question, sir, and very difficult to answer. From the lessons learned, to go back and do it again, to repeat the process, I think this time we would look at existing laws, regulations, policies, and procedures to see where they did not meet the level of what happened, the reality. We would work closer with the State and the locals in sharing responsibility, not to assume their responsibility, not to force ourselves, but to share in implementing those shortfalls that we saw come to light as the real shortfalls.

As an individual, if I was running the situation, I would take these lessons learned and put them into some form of usable, implementable activity that would address—we were building a partnership where we were beginning to lose the yours, mine, and ours syndrome—to me, that would be very necessary to make this part of the lessons learned, is that everybody has a stake in it. Everybody is a taxpayer. We need to work together and try to come to, this is not yours, this is not mine, it is ours and move forward. So that is the biggest thing that I have learned in this whole disaster, is that we need to work closer and stop the concept sometimes of local, State, and Federal Government, but to work more as one unit. I know that is theoretical sounding, but that is a true belief.

Senator CARPER. OK, thank you. Mr. Fontenot.

Mr. FONTENOT. I agree——

Senator CARPER. Again, the question is, looking back at what was learned, how would you ensure that the lessons learned during this exercise were better translated into your particular agency's emergency plan?

Mr. FONTENOT. I agree with what Wayne is saying about the yours, mine, ours concept, and I think that is something that needs to be pushed further, and we tried to do that from day one with the exercise with the contractors. This is a team. This is not an us versus you type of thing.

What would I have done differently? Given the same circumstances that I had back then, I don't know if I could have done anything differently. Rather than getting on the roof and start shouting and then people thought I was crazy and sent me to an asylum, I don't know what else I could have done.

Mr. ST. AMANT. I would listen. My turn?

Senator CARPER. Please. You pronounce your last name St. Amant?

Mr. ST. AMANT. Yes, sir.

Senator CARPER. Thank you.

Mr. ST. AMANT. Please call me Jesse.

Senator CARPER. St. Jesse?

Mr. ST. AMANT. That is fine. [Laughter.]

What we did——"Louisiana Citizens' Awareness and Evacuation Guide." Plaquemines Parish went out and spent some money, and they had every one of these delivered to a person's residential address before the storm hit, about a month before the storm hit. This was produced with Homeland Security funds. It tells people what to do when they have to evacuate. That is one of the results of Hurricane Pam. Public health impacts——

Senator CARPER. We get a lot of things at our home in the mail, and I am sure others do, as well. How do you know whether people, one, read it; two, internalized it, studied it; and three, did anything differently as a result?

Mr. ST. AMANT. That is of no consequence because when we do mandatory evacuation, we put our volunteers and our sheriff's office on the street and the bullhorns, and we tell them, you are under—we don't assume that someone is going to hear the news in an area. We will make it happen. You can never do that in emergency management.

The reason—and let me clear something up, if I may. This agency who put out this dire report or whatever, if it came out at 9 a.m., it is because the sheriff, the parish president, and I were on television telling people to get out of Dodge. BOOT, Be Out Of Town. I don't need somebody from Washington to tell me, as Emergency Preparedness Director, when to advise the parish president or the sheriff what we need to do. They are consummate professionals. They expect me to be one. I expect the government officials, when I advise them to do something, to follow my lead.

The bottom line, sir, or I think the point I am trying to make here, you asked the question, what did we learn? I went back and I took a look at my hurricane plans again——

Senator CARPER. That wasn't my question.

Mr. ST. AMANT. I made some adjustments——

Senator CARPER. No, let me repeat my question again. How would you ensure that the lessons learned during this exercise were better translated into your agency's particular emergency plan? That is my question.

Mr. ST. AMANT. Because I know the vulnerability assessment, we know to get out. Pam, I was there as a subject matter expert as well as the other directors, worked to tell them what they were going to inherit. If you have to tell me what I am faced in a Category 2, 3, or 4 hurricane, they have got the wrong guy for the job. It is my responsibility to prepare that parish to do what is necessary to get out. It was my responsibility under the Pam scenario to advise IEM and the FEMA people what they are going to have to deal with to better prepare them, and to that end, I strongly suggest that they scrap the Stafford Act, which is still governing this emergency instead of a National Disaster Response. But anyway, that is one of the terms that we asked that we learned from Pam.

The other thing was, have a pre-landfall declaration because it gives us the authority and gives us the support necessary to evacuate pre-landfall. Up until this such time, I have to tell you, and this is what was done for Katrina, and it was necessary. Was it done soon enough? History will judge us.

But the fact of the matter is, it is one of the few times in history it is being done. And let me add, if it wasn't for some of my fellow directors and I trying to beg and almost demand that we do a Pam exercise, it wouldn't have happened. It would have gone away. So I just hope someone listens to what is being said, that is all. Sorry for my frustration, sir. It has been a long couple of months, too.

Senator CARPER. Thank you. Ms. Beriwal. The question is probably not entirely appropriate for you because I don't believe you are here representing any particular agency, but would you comment on it nonetheless?

Ms. BERIWAL. I would like to, Senator. Thank you very much. I think that there are certain things that we are taking as lessons learned from this as we go about the country and internationally helping our customers with emergency preparedness.

First of all, I think what we learned is that a scenario-based planning exercise process like Hurricane Pam can be quite effective. Perhaps that is not where your question was going, but I think this is a finding that we have discovered, particularly having the planners and the operational people in the same room so that the operational people have the real-life operational concerns there, and the planners can look at this issue from a wider time scale and a wider geographic scale. It is very effective to have those two groups together.

I think that integrating science and technology like we did into the consequence assessment was very vital. It was very important. It gave a sense of reality to the participants which they carried forward and probably used in Katrina itself. So I think that was a very important lesson learned from that, that we need to integrate our scientific and technological knowledge in this country, which we have a vast amount of, and pour it into these kinds of events.

The third thing, I think, that I would say is that leadership does need to be present, and that is what I would say their role is——

Senator CARPER. I am sorry, what needs to be present?

Ms. BERIWAL. The leadership does need to be present for the different layers of government, and one of their primary roles is in deriving what the outcomes are that they would like to see. I would say that in Hurricane Pam, we worked with all the 13 parishes and we projected 61,290 dead. That was known in Pam. Well, 1,100 dead in Katrina is deemed unacceptable, so we have to define what is acceptable, and that is a role for the elected officials—to decide what is acceptable.

I do think, also, that we need to have emergency management where we can actually take our plans, our doctrines, our training, our exercises, and our equipment and be able to pour that into a single modeling and simulation capability that basically gives us, well, how much protection this is providing because you don't know when you are dealing with hundreds of variables, all of which could have very many different values. Where we are at that point is not known unless you pull this together and are able to quantify protection in some measurable manner.

And third, I would say that our exercises need to be a lot more outcome-based so that when we actually do test a plan, we should be able to see how many people did we save, how many people died, how many people were injured, and could we have done better. It has been one of my maxims since right after September 11—I was actually on the Defense Science Board that looked at intelligence gathering for terrorism, so it hit home closer to me when the events of 9/11 occurred. My maxim to my people was, if we can find a way to save one more person, had we found one person in the Twin Towers that we would have rescued and brought out, we would all as a Nation have been happy. So we would like to find that one additional person that we can save from trauma or death in these kind of events, and we can't do that until we actually have an outcome-based emergency management system.

Senator CARPER. Good. Thank you for a very helpful response. Thank you.

Chairman COLLINS. Thank you, Senator.

The end product from the Hurricane Pam exercise was the Southeast Louisiana Catastrophic Hurricane Plan, and that plan includes an appendix that is entitled, "Transportation, Staging, and Distribution Execution Time Line." The appendix can be found in your exhibit book after Tab B,[1] and I would ask that each of you take a look at it.

On page two, this document indicates that 50 hours before landfall, the plan calls for pre-staging 600 buses and 1,200 drivers. I am going to start with you, Mr. Fairley, and then go across. Was it clear to you whose responsibility it was to stage those buses?

Mr. FAIRLEY. Yes, ma'am, at the time. Working off our normal way of business, local has first-line responsibility, followed by the State, supplemented by the Federal. We came up with an estimated need of around 600 buses. From that, to get to the Federal part, we would have subtracted what the locals would have, followed by what the State would do, and then we would pre-stage or try to pre-stage the remaining. So, yes, ma'am, for me, it was clear based on our normal business activities.

Chairman COLLINS. So the responsibility was first at the local level, then at the State, and then Federal, if requested?

Mr. FAIRLEY. Yes, ma'am, that is the normal procedure. We never tell the Governor what they do or do not need. They will request us to provide those assets.

Chairman COLLINS. And was this plan followed?

Mr. FAIRLEY. Yes, ma'am, it was, but it was not successful.

Chairman COLLINS. Mr. Fontenot, same question for you. Do you think it is clear whose responsibility it was to stage those buses and those drivers?

Mr. FONTENOT. Senator, first, let me say that this happened, this session happened after I left the State of Louisiana.

Chairman COLLINS. Right.

Mr. FONTENOT. I left May 31. This happened in July. But with not being there, yes, it was very clear in my mind whose responsibility evacuations was and whose responsibility that evacuating their citizens was, and it first starts with the local level. Then it goes to the State level, and it is whatever the local level cannot handle, they come to the State and ask for help with, and we try to help them as much as we can. Then whatever we can't help with, we go to the Federal Government to ask for help. It also needs to be pointed out, though, that this is 50 hours pre-landfall——

Chairman COLLINS. Right.

Mr. FONTENOT [continuing]. According to this plan. The Federal Government under the rules that it is under wouldn't come in 50 hours to pre-stage buses for us to have access to at hour 50. They may be pre-staging some assets for later use, but at this point in time, the declaration wouldn't have been made and the Federal

[2] Exhibit B appears in the Appendix on page 80.

Government wouldn't have the authority to turn those buses over to us. However, in my mind, then yes, it was—it is clear by reading this plan. But again, I wasn't there for the discussions so I don't know exactly what discussion went around developing this time line.

Chairman COLLINS. Mr. St. Amant, you are a very experienced emergency management official. Was it clear to you and to the other parishes and to the City of New Orleans who was responsible for staging those buses 50 hours prior to Katrina hitting?

Mr. ST. AMANT. You referenced this booklet, ma'am, and to answer your question, just if you want to read my exact quotation, Exhibit D,[1] the last paragraph, if you don't mind me reading it, and I will answer the question. Jesse, it says, "One message to get to FEMA headquarters is a pre-landfall declaration in a catastrophic situation is very much needed and should be a requirement. There are a lot of people without personal transportation. Therefore, if we don't move people out of New Orleans in an appropriate time, there will be mass casualties. The city at this moment does not have the resources or capability to evacuate these people. Therefore, a pre-landfall declaration is a necessity and a requirement for life and safety."

The issue that I was trying to raise, it was a discussion of all that, not only evacuation resources, etc. We were under the impression that is exactly why we were there, to try to bring out these points of the mass infrastructure lack of capability and the necessary logistics support that would be necessary to move that many people outside of the risk area.

Chairman COLLINS. But there is also a document that is in Exhibit D which contains the notes from what appears to be the final briefing of the Unified Command on July 29, 2005, and it includes a section on transportation. You are listed as a participant in that briefing. And comments that are attributed to Don Day note, "We need to pre-identify the sources for these buses and have them lined up and ready. There are plans to evacuate buses and operators out before the storm, but we are at less than 10 percent done with this transportation planning when you consider the buses and the people."

I am trying to get a sense, given that this plan pretty clearly outlines what needs to be done, why it didn't succeed, and I am wondering if it is because Katrina hit too soon and the planning wasn't completed, or whether there was confusion over who was responsible for what, or whether the State and local entities were simply overwhelmed by a catastrophe of this magnitude. But keep in mind, this is pre-storm, so that is why I am trying to get an understanding. Could you help me better understand this?

Mr. ST. AMANT. Lack of planning, lack of coordination, lack of funding, lack of staff, we can pick any multitude of reasons, excuses why it didn't happen. I remember having the situation where I had three busloads of people ready to get out of a nursing home. This was when I was with the State as an emergency transportation coordinator. I was working at the State Emergency Office. I get a phone call, we have got three busloads of nursing home peo-

[1] Exhibit D appears in the Appendix on page 90.

ple, St. Michael's, and two of the bus drivers got on the plane and went to Atlanta. We had to provide emergency resources to get them out right before we had to close the Interstate down.

People panic, and in this case, when you are looking at the worst nightmare come true, I can understand. It doesn't excuse the fact that we need to be prepared for this, that you need to have plans in place. This is what Pam was trying to accomplish. The fact of the matter is, the lessons learned by these things that we were actually discussing was going to visit us sooner than we anticipated. The purpose for which it was intended was to teach us how and what we needed to do collectively. We recognized the shortcomings. The fact that they were not put in place is only because the lessons learned from Pam were not disseminated down and got to the pubic officials to which it was supposed to serve.

Chairman COLLINS. Thank you.

Ms. Beriwal, my time has expired, so if you could just give me a very brief comment in response to this plan and your assessment of responsibility and whether that was clearly understood.

Ms. BERIWAL. It is my perception that the local authorities were responsible for evacuation of the public and that they would ask for resources and the State would provide resources if necessary, and if States felt that they could not provide those resources, that they would request it from the Federal Government, and that was mostly the discussion.

I would like to clarify one thing, which is that phase one of Hurricane Pam, the four workshops that were held (workshops 1, 1A, 1B), they were all phase one of the planning process where we were going to create the Incident Action Plans. Phase two of Hurricane Pam was expected to be a consolidated plan for the whole area where we would look at the resources and see if those things that we identified in the Incident Action Plan could be implemented. That phase has not started. It is not done, and we are sitting 127 days before the start of the next hurricane season for Southeast Louisiana.

Chairman COLLINS. Thank you. Senator Lieberman.

Senator LIEBERMAN. Thanks, Madam Chairman.

Let me just pick up there with you, Ms. Beriwal. Just so I make sure I understand, the document you—first off, you presented a scenario based on a lot of, I think, very impressive scientific, meteorological data about what the impact of a catastrophic storm in New Orleans might be, and am I correct that in the dramatic and all-too-accurate predictions you made of flooding, of damage to property, of the impact on housing and education, hospitals, health care, etc., that you were assuming the status quo in terms of the government response, in other words, that it inherently showed that something more had to be done because obviously you had a report suggesting that 60,000 people might be killed in such a storm, so no one in government reading that could have said, well, that is OK. Am I understanding what the goal of the Pam exercise was?

Ms. BERIWAL. Let me clarify this by giving an example, Senator.

Senator LIEBERMAN. Yes.

Ms. BERIWAL. The 61,290 deaths were based on the 36 percent evacuation rate from the area, and to come up with the 36 percent evacuation, first, we went through the scientific literature like Jesse mentioned—the UNO study and the Corps of Engineers study on public behavior after storms in Louisiana as well as public opinion surveys——

Senator LIEBERMAN. Can I interrupt a second?

Ms. BERIWAL. Certainly.

Senator LIEBERMAN. Did that estimate also include your evaluation of the existing capacity of State and local agencies to assist in the evacuation?

Ms. BERIWAL. Senator, that is exactly where I was going. Taking those numbers, we went back to the 13 parishes and shared with them our initial numbers for the expected response rate for each parish. We worked with Jesse. We worked with the folks from the other 12 parishes and said, "these are the numbers for your parish. Do you think this is credible or do you have a mechanism to raise this number? Would you like the number to be higher or lower?" because we wanted to make it plausible. So we took our scientific data and then we went back to the parish experts and said, "let us adjust these numbers based on what you think is credible for your parish."

Senator LIEBERMAN. Credible meaning what more you are able to do, or what you are able to do with what you have now?

Ms. BERIWAL. What you are able to do now.

Senator LIEBERMAN. OK.

Ms. BERIWAL. It was really the expectation of your current plan, your current procedures, your current policies, how much evacuation would be expected in your parish. And then we rolled up the numbers based on the feedback from the emergency management directors for the 13 parishes to come up with the 36 percent number.

Senator LIEBERMAN. OK. And then in the plan, which is quite extensive and detailed, what you describe is what the responsibilities of the various agencies, Federal, State, and local, would be to reduce the impact of a catastrophic hurricane, correct?

Ms. BERIWAL. Right.

Senator LIEBERMAN. Now my question is, what happened then? Maybe your contract was over at that point. I don't mean literally over, but what was supposed to happen with the plan because you have now presented a rather dramatic and disastrous set of events—property damage, life lost, over 60,000 people dead as a result of Hurricane Pam projection based on the status quo of what the government was able to do at that time. Then what did—well, what did you expect to happen? Was there any mechanism to implement changes in government so they could achieve better results?

Ms. BERIWAL. Is the question for me, Senator?

Senator LIEBERMAN. Just as a starter, yes.

Ms. BERIWAL. If the State of Louisiana did not have the ability to impact the system, we certainly had a lesser ability as a contractor. So we were tasked to do Hurricane Pam. We did the draft, and then they came forward and asked us to do the subsequent follow-on workshops. We did those, and we were waiting for further direction on where the government wanted us to go.

Senator LIEBERMAN. Were the follow-on workshops, they were after the plan was published? In other words, by my dating, the plan was published in January 2005. The workshops were a little bit later. Were those supposed to focus on what changes the Federal, State, and local governments should enact to try to diminish the impact of this catastrophic hurricane?

Ms. BERIWAL. Actually, the first workshop was in July 2004, and by January 5, 2005, we had done five versions of the planning documents. The second workshop was in November and December 2004.

Senator LIEBERMAN. OK, I have got you. So nothing followed the plan. So I guess I would ask Mr. Fontenot or Mr. St. Amant, what happened with the plan at the State and local government level, and Mr. Fairley to the extent you know what happened at the Federal Government level, because from what we saw, a lot of heroic individual effort by governmental employees at each level of government but also a lot that wasn't done which could have diminished the impact of the storm. Mr. St. Amant, do you want to start? What happened to the plan because obviously there wasn't enough there to mitigate on the status quo the impact of a catastrophic hurricane, which came.

Mr. ST. AMANT. There were certain portions of the plan, in our discussion as a result of our participation, that I was able to bring back and to adjust some of what we did in my jurisdiction of Plaquemines Parish. You have to understand, I was present at all of these planning meetings that I was invited to. My parish president was at one, and he decided that is why I am going to have to go to them, so I can keep him informed——

Senator LIEBERMAN. And is it fair to say that there—I have some sense of you that you were not shy about saying to everyone there that the parish and the City of New Orleans, as far as you could tell, was just not up to dealing with a storm of Katrina-size consequences?

Mr. ST. AMANT. I know these two gentleman quite well and have worked closely. I think you will find that they will verify that, sir. I am just glad that they still invite me to these meetings. I have been known to be just a little bit outspoken because of my passion and concerns for the risk that we have.

Senator LIEBERMAN. My time actually is up, but if you can—Mr. Fontenot or Mr. Fairley, if you would respond to the question. What happened, if anything, to close the gap between the responsibilities the plan gave the State and Federal Government and the reality?

Mr. FONTENOT. I will speak about what happened at the State level up until May 31, 2005, when I left the State.

Senator LIEBERMAN. Yes.

Mr. FONTENOT. Since I was the Chief of Plans, it was my responsibility to try to do something with the plans that came out of Hurricane Pam. One of the things that we recognized right away was that we need to get our State plan in line with the Federal plan, or now the National Response Plan. So I went to my boss at the time, recommended that we—we were at the end of a 4-year planning cycle anyway. We needed to update our State plan anyway. So in the middle of this, why don't we go ahead and just go ahead and do a major revision to the plan to make it more compatible with the National Response Plan, and that was step A.

I concentrated the resources that I had at the time with the State to do that. I thought that was the most important step in the process. That occurred, and as I was leaving State Government, that plan was being implemented in the State. It was being signed off on. It had already been signed off on by all the signatory agencies that had actions in the plan or responsibilities in the plan, and it was being sent to the Governor to be signed off by her, and then I left.

One of my goals that did not get implemented before I left was to then pull the responsible agencies for certain sections of the Pam work and also with the State plan and get them together and talk about and try to do more planning and get them to figure out where the holes were and how to fix the holes. However, since I left, I did not get a chance to do that, but that was one of my personal goals.

Senator LIEBERMAN. I presume, based on what we saw, that between the time you left in May and the hurricane hit in August that not much of what was recommended was accomplished.

Mr. FONTENOT. Sir, I left government. I can't really talk about what they did after I left.

Senator LIEBERMAN. OK. Mr. Fairley, actually, in some ways, I asked you this question last time around. I don't know if you want to add anything.

Mr. FAIRLEY. No, sir. I can add just a little bit. One of the things we were doing in the region was taking what we had developed in these scenarios and taking a very hard look at it and comparing it to what we had in existence as far as our hurricane response checklist and any other plan we had, especially as it went back to the National Response Plan, to make sure that there were no holes, gaps, or bumps in the road that we thought would cause us. That was in formulation. We were also working on requests for additional sessions to go beyond the funding cycle.

What several of us got out of it, Senator, was the enormity of what we had gone through and where we needed to go and that it didn't need to stop. It needed to be permeated out to all Federal agencies, all State agencies, and all local agencies. A lot of Federal agencies have participation and some type of ownership of a lot of things in that area, and we wanted to not necessarily just have a pretty plan, but we wanted to see other agencies be funded to do things and provide offshoot activities. So that is where we were beginning to formulate. Then, unfortunately, the hurricane hit.

Senator LIEBERMAN. I presume that you, Mr. Fairley, Mr. Fontenot, and Mr. St. Amant, all concluded after the Hurricane Pam exercise that your particular level of government, Federal, State, or local, was not adequately prepared to respond to a catastrophic hurricane like Pam or the real Katrina, correct? I am way over my time.

Mr. ST. AMANT. You are absolutely correct, sir. It is our opinion that Federal, State, or local government is not prepared to deal with a catastrophic response.

Senator LIEBERMAN. Correct, Mr. Fontenot?

Mr. FONTENOT. Correct.

Senator LIEBERMAN. Thank you, and thanks, Mr. Fairley. Thanks, Madam Chairman.

Chairman COLLINS. Thank you, and that last answer is why we are here. It is very troubling with the start of the hurricane season only a few months away, I wonder if we have learned the lessons of Katrina much less the lessons of Pam.

I strongly believe that planning and simulations such as Hurricane Pam can greatly strengthen our preparedness and response, and I can't help but think that if Pam had been funded back in the late 1990s or early in 2000, when it was first discussed, and if there had been more of a sense of urgency, more clarity as to who was responsible for what, and better implementation of the plan, that the response to Katrina would have been better. Katrina would have been a natural disaster that was overwhelming and taxed all levels of government regardless, but I can't help but think that evacuation would have gone more smoothly if the plans outlined here had gone into effect, and if there had been a better understanding of the roles of the various entities, and that is why we wanted to learn from you today and get your insights and perspectives.

I very much appreciate your sharing your testimony with us and working with the staff in preparation for this hearing. I hope that we can learn from your experience and that next time we will, in fact, be better prepared. But as each of you has reminded us today, we still have a very long ways to go.

Senator Lieberman.

Senator LIEBERMAN. Thanks, Madam Chairman. I couldn't agree with you more, and in some senses, you go back through this painful history, you see the predictions, the awareness, particularly by people in the region and in the city that this is coming and we are not ready for it. And yet there is some way in which the problem over the horizon doesn't seem quite as real as what you are dealing with today. I guess people just hope and pray that the disaster that everyone says will come one day doesn't come.

But here it came, and we were just there last week, Senator Collins and I and four or five other Members of the Senate, and I must tell you, 4 months after we had been there the first time, a couple weeks after Katrina hit landfall, it was stunning and horrific, really. I have been to areas after natural disasters. I have been to war zones. I was in Kuwait after 1991. I was in Bosnia and Kosovo, and I have just been to Baghdad. I have never seen such extensive damage as I saw in New Orleans and the Gulf Coast generally.

We are motivated by that painful reality and the suffering that people endured because we live in an age when you just have to say, we can't kid ourselves, that there are going to be more natural disasters and, God forbid, there are going to be some unnatural disasters because of the enemies we face in the world today. That is the focus of these investigations.

You have set a foundation in what you did in Hurricane Pam. We are going to try to put it to work so that next time the governments at all levels are more prepared and respond more aggressively to the disaster and the harm will be less, we hope and pray.

Thank you very much.

Chairman COLLINS. Thank you.

This hearing is now adjourned. The hearing record will remain open for 15 days for the submission of additional materials. Thank you very much for your cooperation.

[Whereupon, at 12:15 p.m., the Committee was adjourned.]

APPENDIX

Preparing for a Catastrophe: The Hurricane Pam Exercise

Wayne Fairley
FEMA, RVI

Testimony Before The Senate Committee on Homeland Security and Governmental Affairs

Tuesday, January 24, 2006

Good Morning Madame Chairman and Members of the Committee. I am Wayne Fairley, Response Operations Branch Chief for the Federal Emergency Management Agency (FEMA) Region VI in Denton, Texas. I am presently deployed in response to Hurricane Katrina and I am serving as the Operations Section Chief for the FEMA Joint Field Office in Baton Rouge, LA. I have been with FEMA since 1984 and have been associated with over ninety disasters in these past twenty-two years.

I am honored to appear before you today to talk about "Preparing for a Catastrophe: The Hurricane Pam Exercise" and to follow up on the discussions I have had with your staff on this over the past several weeks as well as answer any questions that you may have. The views expressed in my testimony are my own, and do not necessarily reflect the views of the Department of Homeland Security.

To start off I believe it is best to provide a little historical background on the Hurricane Pam Exercise. As I recall FEMA's goal, based on the 2003 Catastrophic Planning Initiative, was to identify areas of the country that could be vulnerable to catastrophic disasters and, in cooperation with the relevant State and local governments, to:

- Examine projected damages and effects associated with a catastrophic disaster
- Confirm current disaster response capabilities

- Identify anticipated response shortfalls, and
- Initiate comprehensive planning strategies to address the shortfalls

Products developed under the "Catastrophic Planning" initiative were envisioned to include incident-specific response plans for pre-selected geographic regions and disasters, planning templates that could be applied to other areas, and new response contingencies.

In late March, 2004, FEMA HQ notified FEMA RVI that the State of Louisiana had been funded for a catastrophic hurricane plan. Thirteen southeastern Louisiana Parishes (including the City of New Orleans) were selected as the initial geographic focus area for FEMA's "Catastrophic Planning" initiative because of their vulnerability to hurricane disasters. This resulted in the Southeast Louisiana Catastrophic Hurricane Planning Project. This initial concept was to have a draft plan by the end of July, 2004.

The Southeast Louisiana Catastrophic Hurricane Planning Project was designed to bring together responders and decision makers from all levels of government and the American Red Cross to begin analyzing and addressing the overwhelming operational complexities that would be involved in responding to a catastrophic hurricane striking southeast Louisiana. Accepting the fact that only limited funding and time were available, topic specific "planning workshops" using a catastrophic hurricane scenario (Hurricane Pam) to frame the discussions were selected as the best approach for identifying and qualifying the scale of requirements needed to build a plan for responding to a catastrophic hurricane. The results were intended to reveal to the Louisiana Office of Homeland Security and Emergency Preparedness (LOHSEP) and FEMA the shortfalls in existing plans and to begin developing additional plans for catastrophic hurricane response.

Existing plans, strategies, policies, and capabilities were reviewed by LOHSEP before the first workshop. As pre-planning for the first workshop, conducted in July 2004, LOHSEP and Federal representatives identified a list of planning topics, based on those provided by the State of Louisiana, as the most urgent or complex topics needing discussion, including:

- Hurricane Pre-Landfall Issues
- Search and Rescue
- Temporary Medical Care
- Sheltering
- Temporary Housing
- Schools
- Debris

During the first workshop, participants were presented with the catastrophic hypothetical "Hurricane Pam" disaster scenario to frame discussions and then divided into breakout groups by responsibilities and topic for detailed discussions. The breakout groups identified operational concerns in each topical area, addressed issues, and drafted plans for dealing with the identified concerns. To address other urgent subtopics that emerged during the discussions, additional breakout groups were established. The following additional subtopics were discussed:

- Access Control and Reentry
- Billeting of Federal Response Workers

- Distribution of Ice, Water and Power
- Donations Management
- External Affairs
- Hazardous Materials
- Transition from Rescue to Temporary Housing
- Unwatering of Levee Enclosed Areas

It became clear after the first workshop that a series of workshop cycles would be necessary to address the full range of complex response and recovery concerns associated with a catastrophic event. Additional workshops were held in November 2004, July 2005, and August 2005 to provide further input for topics. Topics selected for further discussion during the subsequent workshops included the following:

November 2004 Workshop Topics

- Sheltering
- Temporary Housing
- Temporary Medical Care

July 2005 Workshop Topics

- Transportation, Staging, and Distribution of Critical Resources
- Temporary Housing

August 2005 Workshop Topic

- Temporary Medical Care

The goal of the Southeast Louisiana Catastrophic Hurricane Planning Project was to begin addressing immediate, intermediate, and long-term needs; create plans immediately usable by planners and responders in the field; and seed the eventual development of a comprehensive and systematic operational plan. The ultimate goal is for the concepts identified in the Southeast Louisiana Catastrophic Hurricane Planning Project to be integrated into a final catastrophic plan. The project did not result in a catastrophic planning document per se, but rather a framework for developing such a plan.

My participation in the process included working with a LOSHEP counterpart as a member of the Steering Committee. That involves project management, workshops design and participation, budgeting, and HQ and contractor interface. At the workshops this included monitoring of workshop sessions, providing FEMA law, regulation, and policy information, dispute resolution, and overall directional guidance in meeting workshop objectives.

Participation included LOHSEP, State Emergency Support Functions (ESF), local Emergency Management staff from the thirteen southeast parishes, FEMA RVI, FEMA HQ, RVI ESFs, other Federal agencies as requested, and private industry partners.

Areas of responsibility were assigned in the workshops according to existing state and Federal laws, regulation, policies, procedures, and plans. No planning effort was made to re-create or modify any existing authorities. Directed or institutional agency authority on any given subject area was only reviewed and used as guidance by the planning session participants; however, participants were able to comment and provide opinions on existing state and Federal laws, regulations, policies, procedures, and plans and the possible need for changes. Two such State Plans included the Louisiana Hurricane Evacuation Plan and the Louisiana Sheltering Plan.

These planning sessions laid the ground work for future detailed subject plans. They identified the primary areas of concern by the local, state and Federal agencies and began the process of identifying who would address these areas and how they would be addressed. These sessions brought together persons responsible for the implementation of emergency management from all levels of government and helped lay a ground work of cooperation that had never existed.

Future intentions were to include continued subject specific sessions. Some topics were to be expanded. Some topics would be added. Some topics would only be maintained with updated data. It was our hope that the plan would not end or become stagnant but would continue to be a fresh and growing plan that included new data and innovative ideas. It was also hoped that the new formed working spirit between the locals, state, Federal and private industry would continue to grow and lead to a concept of "ours" versus yours or mine.

Although the catastrophic planning process has been interrupted by the impacts of Hurricanes Katrina and Rita, the workshops and planning process – knowledge of inter-jurisdictional relationships and capabilities, identification of issues, and rudimentary concepts for handling the consequences – have been beneficial to hurricane response activities.

I know that this Committee and others are very concerned about what occurred as a result of Hurricane Katrina in Louisiana and I want to assure this committee that all of my fellow employees at FEMA are as well. I want to assist this committee in any way I can in ensuring that what occurred never happens again. I want to thank the Members of the Committee for their past support of FEMA and I appreciate the opportunity to testify before you today. I would now be pleased to answer any questions you may have.

Preparing for a Catastrophe: The Hurricane Pam Exercise

Statement before the Senate Homeland Security and Governmental Affairs Committee

by

Sean R. Fontenot
Former Chief, Planning Division
Former Chief, Preparedness Division
Louisiana Office of Homeland Security and Emergency Preparedness

January 24, 2006

Dirksen Senate Office Building

I would like to thank the committee for inviting me today to speak on the events of the planning exercise known as "Hurricane Pam." As part of the Southeast Louisiana Catastrophic Planning Project, with this event we began the process of trying to fully understand and prepare for the effects of a catastrophic hurricane in Southeast Louisiana. However, before getting into the events of the exercise, I would like to speak first about the events that led up to Hurricane Pam.

Background

In 1998, the Louisiana Office of Emergency Preparedness (now known as the Louisiana Office of Homeland Security and Emergency Preparedness or LOHSEP), realized after Hurricane Georges that more planning was needed for post-landfall consequences after a major hurricane. A working group was convened consisting of Federal, State, and local participants to brainstorm the issues that Louisiana would be facing if a Category 3 or higher storm ever hit Southeast Louisiana. The work of this meeting and a later meeting held in New Orleans in 1999 was the development of a white paper, which outlined a planning proposal that was submitted to FEMA in August of 2000. The paper requested FEMA's help in planning and preparing for a catastrophic hurricane that could hit Southeast Louisiana. A year later, in August of 2001, a second letter was sent to FEMA Headquarters through the FEMA Region VI office, requesting that the same action be taken.

Later in August 2001, FEMA Headquarters awarded a contract to URS Corporation for catastrophic planning support. However, due to the events of September 11, 2001, there were many delays that occurred in this planning process. In December 2001, a kick-off organizational meeting was held in New Orleans with FEMA Headquarters, FEMA Region VI, and State participants, to organize a process to use the URS contract that was awarded earlier that year.

At this December meeting, a committee structure was developed and a plan was formulated for catastrophic planning. This included a budget that the State and FEMA Region VI would need to carry out the planning process. In January 2002, FEMA Headquarters informed the State and Region VI that there would be no further funding for this project, due to budget shortfalls. The funding already in place could be used by the contractor to do what it could on its own, which included identifying large tracts of land that could potentially be used for temporary housing setup.

Following Hurricane Lili, the process was revived again for a short period in December 2002. FEMA Headquarters sent a representative to help LOHSEP reformulate the plans that we had developed the year prior, and to establish a budget. This request was sent back through the chain to FEMA Headquarters, proposing to use mitigation funding, and once again the request was turned down.

In September of 2003, there was a conference call with FEMA Region VI and FEMA Headquarters to discuss catastrophic planning. This led to a meeting on November 18, 2003, in New Orleans on this subject. Attending this meeting was a representative from the President's Homeland Security Advisory Council. At the meeting LOHSEP and

FEMA Region VI briefed the need for catastrophic planning, and he was astonished that as of that date we had not completed this type of plan, and promised to do what he could to help us get funding for this planning process.

This brings us to the Southeast Louisiana Catastrophic Planning project. On March 17, 2004, FEMA Headquarters called FEMA Region VI and the State of Louisiana and informed us that FEMA HQ had funding for catastrophic planning for Southeast Louisiana. The very next day, on March 18, 2004, LOHSEP and FEMA Region VI quickly organized a Unified Command, consisting of the leadership of LOHSEP and FEMA Region VI. A concept was presented to the Unified Command by the lead planner from FEMA Region VI, which was quickly adopted and approved by the Unified Command. At this time, LOHSEP and FEMA Region VI formed a Steering Committee to organize the planning event. This Committee consisted of me representing LOHSEP and a representative from FEMA Region VI. (Later, a representative from FEMA Headquarters was added to the Committee as well.) The very next week, an initial meeting was held in Baton Rouge to discuss this plan of action, which led to the creation of Hurricane Pam. On April 7, 2004, a meeting was held during the National Hurricane Conference in Orlando, Florida, to discuss the concept with representatives from FEMA Headquarters, and to request that FEMA Headquarters find a contractor that could support this planning process.

On May 19, 2004, I was at FEMA Region VI in Denton, Texas, working on the details of the proposed exercise when we were informed by FEMA Headquarters that they intended to award a contract to IEM, Inc. to support this planning project.

FEMA Headquarters faxed the proposed Statement of Work (SOW) to FEMA Region VI for the contract that was to be awarded, and said that IEM would be there to help the State and the Region to get the planning done in the timeframe we had been given. The SOW was not exactly what LOHSEP and FEMA Region VI told FEMA Headquarters that we wanted. However, Headquarters assured us that the SOW was written flexibly in order to get the contract in place quickly, so that we could get started with whatever planning needed to be done in the timeframe that was allotted. We had to be through by September 30, the end of the Federal fiscal year. When we first started looking at this concept in March, the exercise was scheduled for June of 2004. We had to keep pushing the start of the exercise back until a contractor was in place to support the project. However we also knew that this planning event had to take place before August because the height of the hurricane season for the Gulf of Mexico is typically August and September.

From the word "Go," it was understood that this wasn't going to be a typical exercise. In fact, when the concept was first given to me that we were going to have an exercise to develop a plan, I immediately disagreed. Usually, you write a plan and then have an exercise. However, when it was explained to me that we were taking an exercise scenario which generated real consequences and real data and bringing operational level people in so they could make decisions using that real data and consequences, which would then drive the writing of the plan, I quickly got on board. I championed the fact that we were using operational people to write the plan; because there are too many times a plan is written without taking the operational aspects into account and this leads to non-usable plans.

Also, it needs to be pointed out that all levels of government had existing plans prior to this exercise. This planning exercise was developed to work on a "bridging document" that would form a bridge from the local to State to Federal levels.

In the initial concept, the exercise was set up to last for 14 days, with all participants involved for the full two weeks. However, FEMA Headquarters said the price for an event that long was too high. So, the exercise was reduced to eight days (three days pre-landfall and five days post-landfall); only the five days of post-landfall discussion included all participants. (The Planning and Steering Committees met the day before the exercise started as well as the day after the exercise ended.)

We started the exercise with three days of pre-landfall discussions with just FEMA Headquarters, Region VI, and the State, to set the tone for the rest of the week. These discussions were very focused on what the Federal Government and the State would be doing pre-landfall: what the State would be requesting, and what the Federal Government would be staging. FEMA Region VI looked to LOHSEP to choose the topics that the exercise was going to concentrate on, however all topics were agreed upon by both FEMA Region VI and LOHSEP before we moved forward. It was the belief of LOHSEP's decision-makers (myself included) that we needed to focus on human needs first. That's how most of the topics for the exercise were chosen: Search and Rescue, Sheltering, Medical Care, Long-term Housing, Schools—with the one exception being Debris Management.

We initially met with the exercise contractor, IEM, at FEMA Region VI in Denton, Texas, on May 20. By that point, Region VI and the State had already spent two months preparing the details of how we wanted to do things, and we presented this concept to the contractor at that meeting. I pointed out, and FEMA Region VI agreed, that we had to work as a team and stick to our game plan to get this event accomplished in the time period available. We had only 53 days to put together something that would normally take six months to a year to plan. And we couldn't push it back any further because August and September are the "hot" months for hurricanes in the Gulf. (We were already pushing it by holding the exercise in July.)

We tried to involve local emergency managers as much as we could. For instance, when IEM developed a set of consequence estimates, the Planning Committee (LOHSEP, FEMA Region VI, and IEM) would meet with or poll local emergency managers to include them in the planning process from the beginning. Because of the tight timeframe, they were not as involved as we would have liked, but we tried to keep them involved as much as time would allow.

Participation in and Evaluation of the Exercise

The Hurricane Pam exercise ran from July 16-23. On a typical day of the main exercise, we had six breakout rooms which had the same assigned topics for the entire week. Then we had three action rooms, which were assigned topics on a day-to-day basis. Each day, the breakout rooms were responsible for writing a certain portion of the action plan based on the template that had been agreed upon by FEMA Region VI and LOHSEP prior to the event. The contractor had a facilitator and a recorder in every room to make sure that the room completed its task for the day. Also assigned to each room were a Federal and a State lead who were responsible for briefing the Unified Command on a day-to-day basis. This process lasted for five days.

Each day of the main exercise started with a briefing, led by me, to the entire 300+ participants, in the Operations Center at LOHSEP. This briefing included a presentation by a representative of the National Weather Service (Southern Region) of the day's simulated weather events from the fictitious Hurricane Pam.

After the weather briefing, a briefing on the scenario consequences that resulted from the weather conditions was given by the contractor, IEM. Then the operations officers from LOHSEP and FEMA Region VI would take the action room topics determined by the Unified Command for the day and assign agencies to send representatives to those action rooms. Then, the main body would break up and participants would go to their assigned breakout and action rooms.

FEMA Region VI and LOHSEP expected that the action rooms would only produce the beginnings or the framework of a plan that would have to be fleshed out later; the breakout rooms, since they had five days to work on the plan, were expected to produce more of a complete plan. However, we also knew that the breakout rooms would not develop a 100% answer within the five days that we had to work with either.

I was the State lead for the day-to-day decision making for the design and execution of the planning exercise, with guidance and input from my Assistant Director. I was the main point of contact for the State in dealing with the contractor and FEMA. I worked closely with them during the exercise, making sure that we stayed on schedule and that deliverables were being met in the breakout rooms. I was also responsible for working with the Unified Command to determine the topics for the action rooms on a day-to-day basis, and making sure that any questions the Unified Command had were getting answered.

During the breakout sessions, I tried to visit each breakout room to make sure that they were staying on track and to answer any questions they might have had about the scenario or what the responsibilities of the day were. During this time, I also worked with the contractor to help resolve any issues that came up during the course of the week; I ensured that the briefings to the Unified Command were happening on schedule; and I worked with the FEMA and the State operations officers to make sure that the right people were assigned to the appropriate rooms and topics. I also dealt with the many routine logistical issues associated with a weeklong gathering of more than 300 people.

On the last day of the exercise, a meeting was held with representatives from FEMA Headquarters, FEMA Region VI, the State, and IEM—basically, the Planning and Steering Committees of the exercise. We talked about the next steps and delivery schedules. We felt the need for follow-on workshops, but we did not come to any official decisions that day about what those follow-on workshops would cover.

Since this was not a standard exercise, there was no formal evaluation process. As I mentioned previously, this was an exercise that was designed to develop a plan, not to test a plan. In my opinion, the exercise was very successful—not because we developed the perfect plans, but because it brought operational-level players to the table to begin the planning process. We never expected to come up with a 100% solution. It was always felt that if we had a 70% start, that we would be successful. The scenario-based planning exercise, in my opinion, has produced the foundation for a very successful plan. However, due to funding and time constraints, we had to be very selective about the topics that were covered during the main exercise and during the follow-on exercises.

Understanding of the Scenario

The Planning Committee—which consisted of FEMA Region VI representatives, me as the State representative, and the contractor—developed a scenario to show that it did not take a Category 4 or 5 hurricane to cause catastrophic effects in Southeast Louisiana. I wanted to show that a Category 3 on the right track could cause these catastrophic events to occur. The storm was based on a Hurricane Georges scenario, except in this case, the hurricane didn't turn toward Mississippi (as Georges did), but hit Louisiana directly.

The National Weather Service (Southern Region) helped in the development of the weather scenario for Hurricane Pam. I gave them the overall parameters: I wanted a slow-moving Category 3 hurricane that overtopped the levees in New Orleans, coming from the southeast, moving northwest. The National Weather Service, working with other NOAA partners, came up with the exact track and the characteristics of the storm. The overtopping of the levees was included to cause the catastrophic flooding conditions from the storm surge.

The contractor was responsible for development of the consequences based on the storm scenario that the National Weather Service developed—fatalities, communications outages, energy outages, cubic yards of debris, home damages, etc. All consequences were reviewed by the Planning Committee and the Unified Command; in addition, certain consequences were reviewed by the parish emergency management officials. This was to ensure the believability of the consequences and to get buy-in from the local emergency management officials. We knew that if the consequences weren't believable, then the focus of the players during the exercise would be on disputing the consequence numbers and not on developing the plans.

Although the Planning Committee felt that we were very successful in getting buy-in from most parties, there were still some side conversations discussing the consequences. We tried to keep these to a minimum in order to stay focused on the task at hand—developing the bridging document. For instance, there were some discussions about the projected number of fatalities. Even though the parishes agreed with the numbers, some Federal participants from FEMA Regions IV and VI said during the exercise that they felt the projections were too high. (The numbers were based on a combination of a US Army Corps of Engineers Study, historical evacuation rates from Hurricane Georges, and ground-truthing conducted with parishes and the State.) Also, there were questions by some participants about the projected flood depths of the waters that were based on the approved model (the Sea, Lake and Overland Surges from Hurricanes, or SLOSH model); however, further analysis showed that other models produced similar numbers.

Areas of Agreement Regarding Responsibilities and Response Times by Parish, State, and Federal Agencies

It has always been common knowledge within emergency management that the locals and the States would have to be prepared to sustain themselves for a period of 48-72 hours before they could expect major Federal resources to be on the scene. It has always been taught in emergency management that the Federal Government is not a first responder. There are some exceptions to this with the Coast Guard and other agencies, but for the most part, the Federal Government is not (and in my opinion should not be) a first responder.

As far as the Hurricane Pam exercise is concerned, since I was not in every breakout room 100% of the time, I can't speak specifically to verbal promises that anyone made to anyone else in the rooms. The official results of those discussions in the breakout rooms were captured within the Southeast Louisiana Catastrophic Hurricane Plan. After the exercise was over, every participant received a copy of the plans that came out of the exercise. Again, we understood that we just had the beginning of the answer, and more collective work by the Federal, State, and local participants was required to resolve all the issues that were raised.

Planning LOHSEP Undertook to Incorporate Lessons Learned from the Exercise.

One of the primary things LOHSEP recognized at the conclusion of the exercise was that we needed to update our State plan to reflect the Federal Response Plan (now the National Response Plan).

The State plan is on a four-year planning cycle, and we were at the end of that cycle, so it was time for an update. We also knew that the Federal Response Plan was going through a major revision to become the National Response Plan. As the Planning Chief, I advised the Deputy Director for Emergency Management that we should wait for a final draft version of the National Response Plan to come out before we updated our plan, so that we didn't have to update it twice. We started that process in late 2004.

Essentially, we changed our State plan from a functional format to the Emergency Support Function (ESF) format, including the 15 ESFs associated with the National Response Plan. In the process of updating the plan, all functional areas, with the exception of one, remained with the existing State agency that had been responsible for that function prior to the plan update. The one function that changed was transportation. In our old plan, transportation was the responsibility of the National Guard. In the new plan, this function was transferred to the Louisiana Department of Transportation and Development. The final result of the plan update was that we had equivalent agencies at the State level talking to their Federal counterparts (in this case, Louisiana DOTD talking to the Federal Department of Transportation).

Issues Identified in July 2004 that Required Further Discussion

As I mentioned, it was generally understood at the end of the exercise that we did not have anything close to a 100% answer for a catastrophic event. It was very evident at the end of the week that there were certain groups that still needed to meet for further discussions and dialogue on their plans. During the course of the exercise, it was suggested that we continue this process with some follow-on workshops. It was agreed by the rest of the Unified Command, FEMA Region VI, and the State that this would be a good thing. This led to the development of the follow-on workshops, which I was the lead for, until I left LOHSEP on May 31, 2005. We had the first of these follow-on workshops in late November/early December 2004.

Initially, LOHSEP proposed to FEMA to have a second major planning event like the first Hurricane Pam exercise to focus on some of the areas that we did not get to cover during the first exercise. These areas included Transportation, Communications, Feeding, Security, Financial issues (Banking, Insurance, Cash flow to affected individuals, Postal delivery of financial documents, etc.), Personal records, and Missing Persons/Family Reunification. The Unified Command also felt that External Affairs/Public Relations, which was covered at the first Hurricane Pam exercise, needed additional work.

However, in the meantime, the Steering Committee scheduled a follow-on workshop for late November/early December to focus on continued planning for Sheltering, Housing, and Temporary Medical. These workshop participants took the plans they created during the initial Pam exercise and tried to expand them and fill in the holes that they knew were there by bringing additional information to the table.

During this whole process, there were many delays in the execution of the follow-on planning exercises due to funding issues. These issues were mainly dealing with Federal funding for Federal travel. FEMA funded the State and (to my knowledge) the contractor with end-of-year funds. However, due to Federal budget rules, they could not pay for Federal travel with the same funding, which led to delays for the follow-on workshops.

It became clear after the first follow-on workshop that there would not be another large exercise, due to funding. Therefore, FEMA Region VI and LOHSEP decided to use the 2nd follow-on workshop to focus primarily on the Transportation, Staging, and Distribution of Critical Resources plan, because it affected all the other plans one way or another. There was also further discussion of Temporary Housing. In addition, there was supposed to be a follow-on discussion of Temporary Medical Care; however, due to some scheduling conflicts, it wasn't included in the second follow-on workshop. It is my understanding that the medical experts met in a separate workshop in the summer of 2005, but I was no longer with the State at that point, so I don't know what came out of that workshop. Although I came to work for the Hurricane Pam exercise contractor

(IEM), I recused myself from any Hurricane Pam follow-on activities due to my previous State responsibilities, in line with counsel I received from the Louisiana State Ethics Board.

In discussions at Hurricane Pam and the follow-on workshops, one of the things the participants realized in the Temporary Housing arena (and in some of the other instances) was that the Stafford Act was not capable of handling a disaster of this magnitude. Some of us felt that changes needed to be made to FEMA policies regarding some of these issues. Another example was the rules for the Public Assistance (PA) Program as they related to schools. There was a need to help schools outside the affected area because their costs doubled when their student populations doubled overnight. There are no rules in the PA program covering that issue, to my knowledge. We didn't have time to get into the discussion of how to change these policies and rules; the first step was to identify the issues. I always thought this was a discussion we needed to have down the road. But since I left the State and this planning process, I was no longer in that loop.

In spite of the funding, scheduling, and policy challenges we faced with the Southeast Louisiana Catastrophic Planning Project, I felt that we had started down the right path. We still had a way to go, but we were headed in the right direction. It is my opinion that scenario based planning activities, like Hurricane Pam, are the way to go when trying to formulate plans to deal with catastrophic events. The realism that is brought to the table during these events really makes the planning feel more urgent.

OPENING REMARKS
Jesse St. Amant, Director
Plaquemines Parish
Office of Homeland Security & Emergency Preparedness

Before the
United States Senate Committee on Homeland Security & Government Affairs
January 24, 2006

- **Hurricane Pam scenario**
    - **Designed to demonstrate response and recovery ONLY,** to the federal response agencies the potential catastrophic result for
    - State agencies and local emergency directors provided the local scenario effects and considerations for response and recovery efforts
    - **Demonstrated - beyond the state and local capability**: FEMA should have been prepared to support them

- **Lack of interoperable communications** make it impossible to coordinate response

- **Pre-landfall declaration of emergency** should be standard for State and Local evacuations
    - The Federal Government needs to be proactive not reactive
    - Plaquemines Parish is an example of proactive planning

- **Lack of unity of command** within FEMA hurt response activity

- **FEMA organization:** Should be restored to independent agency with direct access to President

- **FEMA policy is inconsistent, ambiguous, and unclear**
    - Debris removal, channel, harbor and waterway clearing – boats & debris but not silt
    - Temporary housing programs

- **Stafford Act is inadequate** to respond to catastrophic events
    - Federal coordinating officer needs authority to better direct use of Federal resources

- **FEMA slow in providing Public Assistance funding to local governments**

- **Crisis and consequence management**
    - Short Term Response vs. Long Term Recovery
        - Houston housing evacuees – crisis response with consequences, i.e. housing, schools, police

- **Military response coordination**: While necessary should be coordinated through the State and Local direction

- **Planning shortfalls**: Federal Government (FEMA) and state agencies and local governments have known about potential catastrophic results for 20 years w/no substantial planning effort, with the exception of Response 95.

- **Roads and transportation infrastructure** to expedite evacuation

IEM

Preparing for a Catastrophe: The Hurricane Pam Exercise

Statement before the Senate Homeland Security and Governmental Affairs Committee
Madhu Beriwal, President and CEO, IEM
January 24, 2006

Chairman Collins, Senator Lieberman, and members of the committee, I appreciate the opportunity to testify today on Hurricane Pam and its role in catastrophic planning and preparedness. My name is Madhu Beriwal and I am the President of IEM. IEM was the prime contractor for the Southeast Louisiana Catastrophic Hurricane planning workshops generally referred to as "Hurricane Pam."

First, I would like to mention my background and that of my company. I have 25 years of experience in emergency management and homeland security, and the application of information technology to these areas. I began my career with the State of Louisiana, working in floodplain management and hurricane emergency preparedness for New Orleans and the surrounding areas. In 1985, I founded IEM, a disaster consulting firm dedicated to keeping people safe—at home, at work, and on the battlefield. We have worked closely with many federal, state, and local organizations to improve preparedness for hazards ranging from natural disasters to those involving chemical, biological, radiological, and nuclear threats. We have worked with the Chemical Stockpile Emergency Preparedness Program for 15 years, helping to improve preparedness for accidents involving lethal chemical weapons stored at 8 sites in the continental U.S. For the Department of Homeland Security's Office of Grants & Training Office (formerly the Office of Domestic Preparedness), we are providing state and local agencies with technical assistance in developing prevention and preparedness plans and procedures. Over the last 21 years, IEM has received numerous awards, including the National Reinventing Government Hammer Award, the James S. Cogswell Award, a Profiles in Innovation Award in Emergency Preparedness & Response Excellence from GOVSEC, and, in 2004, the first DHS Emergency Preparedness and Response Under Secretary's Award for Superior Service awarded to a contractor.

At IEM, we base our work on sound science to provide objective solutions that support difficult decisions about preparedness and protection. We strive hard to create solutions that allow all stakeholders to collaborate effectively. IEM is known as an "honest broker." Organizations like the Department of Homeland Security and Department of Defense routinely rely on our company to function as an independent evaluator of emergency planning strategies and tools. We are called upon to perform independent validation and verification for mission-critical systems and to conduct comparative analyses of various protective options, often in politically charged environments. But most importantly, we at IEM are passionate about designing and producing effective outcomes that help our customers and stakeholders keep people safe.

Hurricane Pam

IEM, with a team of three subcontractors, competed for and received a FEMA contract for catastrophic planning for Southeast Louisiana and the New Madrid Seismic Zone on May 24, 2004. Because of the urgency associated with the project, IEM personnel met with FEMA Region VI officials and State officials from the Louisiana Office of Homeland Security and Emergency Preparedness (LOHSEP) within 18 hours after verbal notice of contract award. At this initial meeting, the overall purpose of the project was presented by FEMA Region VI and LOHSEP. The goal was to develop a functional, scenario-based exercise that would drive the writing of Incident Action Plans and build the foundation for Functional Plans. Ultimately, the project was intended to create a "bridging document" between local and state plans and the National Response Plan. The first "Hurricane Pam" workshop was held 53 days after contract award. As the planning proceeded, it became evident that multiple workshops would be required.

Hurricane Pam was an innovative concept that combined two facets of emergency management: planning and exercises. Traditionally, a small group of planners first develops an emergency plan, and then training is provided to those who will execute the plan. The plan is then exercised to identify gaps, omissions, and areas for improvement. Plans cover a wide range of hazards; hazard-specific plans cover from small-scale to large-scale of the same hazard. Exercises are typically scenario-based and consider a specific event that could happen. The full process takes time—plan development generally takes 6 to 18 months, and training on the plan may require 6 to 12 months. Then small-scale exercises lead to larger exercises where hundreds of personnel and dozens of agencies participate to test the generic plan against a specific scenario. Planning, execution, and evaluation of exercises will generally add 18 to 24 months to the process. The full process of planning, training, and exercising can take 2 ½ to 4 ½ years to complete. Hurricane Pam was a "planning exercise" designed to develop usable information in a much shorter timeframe.

In contrast to the traditional planning and exercise process, Hurricane Pam allowed both planners and operational personnel to collaborate in developing a plan based on a specific scenario. Hurricane Pam focused on developing plans for a specific catastrophic hurricane striking Louisiana. Thus, it was both a planning workshop and a scenario-based exercise. All 13 parishes and most of the 20+ Louisiana State agencies, and 15 federal agencies involved in Hurricane Pam had emergency plans or procedures and many of them had hurricane plans. **Hurricane Pam was designed to bring planners and decision-makers together from all levels so they could begin to grapple collectively with response issues for a catastrophic event and start the process of reviewing and reconciling their existing emergency plans.** The intent of Hurricane Pam was to produce the preliminary "bridging document" addressing catastrophic hurricane response between state and local plans and the National Response Plan. The Hurricane Pam documents were designed to serve as a foundation for more detailed catastrophic planning in the future, and to provide the architecture for an integrated catastrophic plan. (Note: The National Response Plan was not finalized until early December 2004, while the Phase 1A workshop was being conducted, and 5 months after the initial workshop was conducted.)

The 300+ workshop participants at the Hurricane Pam workshop in July 2004 were provided with a catastrophic hurricane scenario, a set of consequences that would result from that scenario, and assumptions designed to stress the emergency management system and force thinking on critical planning topics. In addition, they received a copy of Louisiana's Emergency Operations Plan, 12 parish emergency operations plans, the City of New Orleans' Comprehensive Emergency Management Plan, and emergency plans and related documents from Louisiana Department of Transportation, Louisiana State Police, the US Army Corps of Engineers, and FEMA Region VI.

To create catastrophic conditions, Hurricane Pam was modeled as a strong, slow-moving Category 3 storm preceded by 20 inches of rain, spawning tornadoes and storm surge, and resulting in 10 to 20 feet of water within the City of New Orleans. A slower hurricane builds a higher head of storm surge and is more catastrophic, as historically 9 of 10 storm-related deaths are due to drowning. (Note that according to a December 20, 2005, report by the National Hurricane Center, Hurricane Katrina was a faster-moving Category 3 storm when it reached the Louisiana gulf coast.)

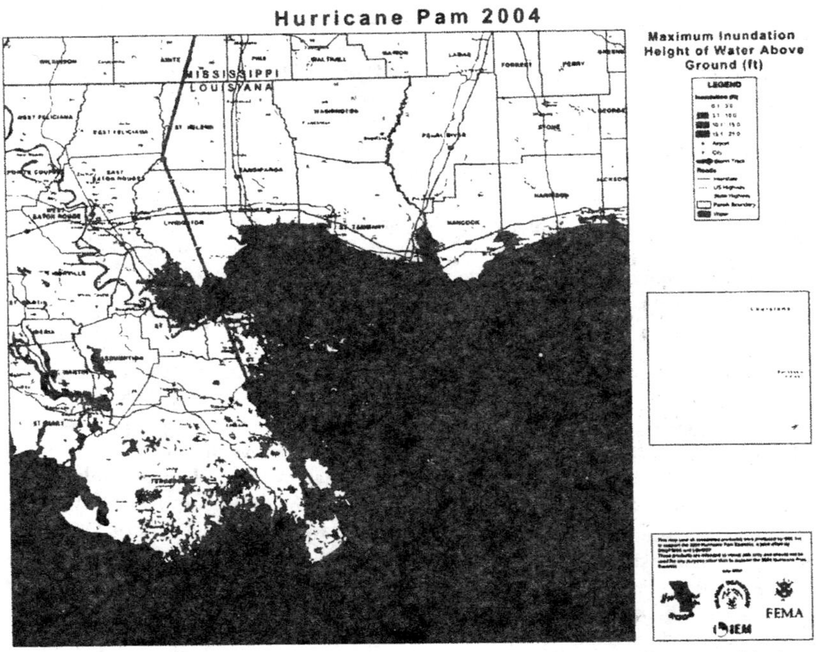

Figure 1: Map from "Hurricane Pam" illustrating maximum height of water above ground in the affected areas (storm surge and rainfall).

To present participants with a realistic situation and to provide context for the event for which they were planning, IEM projected a detailed list of consequences that would be expected from the hypothetical Hurricane Pam. For example, IEM calculated that 55,000 people would be in public shelters outside Southeast Louisiana prior to landfall, more than 500 miles of major roads would be flooded by the storm, 1.1 million residents of Southeast Louisiana would be displaced, and that 80% of the structures in the 13 parish area would be affected by wind and flooding, varying from minor wind damage to total structural collapse. (*See Table 1 for a list of other projected consequences.*) Affecting more than 12,000 square miles in Louisiana and over 1.9 million people (residents and tourists), Hurricane Pam presented a complex web of topics and missions to be addressed in the planning for such an event.

Hurricane Pam was designed to be a series of workshops, conducted as "phases," focusing primarily on post-landfall response issues. Before August 29, 2005, when Hurricane Katrina struck, four workshops had been completed. At the first workshop in July 2004, there were over 300 officials from Federal, state, local, and voluntary organizations in attendance. The second and third workshops had over 100 officials each. The last workshop, completed four days before Hurricane Katrina struck, had about 80 officials. All total, there were more than 350 unique attendees from Federal, State, local, and voluntary organizations. The atmosphere in these workshops was intense, focused, and dedicated. Participants knew that they were facing a real threat, as articulated in the detailed scenario. There were intense discussions on strategies—open, participatory, and creative brainstorming on how best to protect a deeply vulnerable region from a massive catastrophe.

Hurricane Pam Phase 1

The first Hurricane Pam workshop was conducted in Baton Rouge on July 16-23, 2004; approximately 300 federal, state, and local officials attended. These included representatives from FEMA, over 20 Louisiana State agencies and organizations, 13 parishes, the National Weather Service, over 15 federal departments and agencies staffing the Emergency Support Functions (ESF), Emergency Management Assistance Compact (EMAC) participants, volunteer agencies, and neighboring states of Mississippi and Arkansas.

The workshop was organized by topics determined by FEMA and the State of Louisiana prior to the workshop. These were: Hurricane Pre-Landfall, Schools, Search and Rescue, Sheltering, Temporary Housing, Temporary Medical Care, and Debris Management. Based on the consequences presented to them, the 300+ participants in Hurricane Pam Phase 1 began developing action plans for these topics. As the week progressed, it was evident that additional topics urgently needed to be addressed. These topics were added and covered over one or more days of the eight-day workshop. These additional topics were Billeting of Emergency Response Personnel; Hazardous Materials; Power, Water and Ice Distribution; Access Control and Re-entry; Unwatering; External Affairs; Transport from Water to Shelter; and Volunteer and Donations Management. Many other topics could not be addressed within the 8-day workshop and were deferred for future workshops.

Use of topics rather than Emergency Support Functions (ESFs) allowed cross-ESF thinking on each topic. Since it was evident that many issues were interrelated, such as

Search and Rescue, Sheltering, and Temporary Medical Care, joint-topic committees were formed during Hurricane Pam to address these issues together.

At the end of Phase 1, it was apparent that the complexity of particular topics necessitated additional planning attention. Subsequent workshops were held to focus on these issues. In particular, Sheltering, Temporary Housing and Temporary Medical Care issues were all addressed more than once, some three times over the four workshops.

Hurricane Pam Phase 1A

Hurricane Pam 1A was held November 29-December 3, 2004, in New Orleans. Attended by approximately 100 officials, this workshop focused again on Temporary Housing, Temporary Medical, and Sheltering.

Hurricane Pam Phase 1B

Hurricane Pam Phase 1B was conducted July 25–29, 2005, in New Orleans, and was attended by 100 officials. This workshop focused again on Temporary Housing and addressed a new topic: Transportation, Staging, and Distribution of Critical Resources.

Hurricane Pam Temporary Medical Care Workshop

On August 23–24, 2005, the Temporary Medical Care Supplementary Planning Workshop was held in Carville, Louisiana. It was attended by approximately 80 officials. This was the third workshop that addressed medical issues.

Many topics were deferred for future workshops were not addressed before Hurricane Katrina struck. These include: Security, Command and Control, Feeding, Communications, Continuity of Government Operations, Banking and Finances, Missing Persons and Family Reunification, Personal Records, and Recovery and Rebuilding of Infrastructure.

Hurricane Pam Results

There is a maxim in warfighting "No plan survives first contact with the enemy." There is another in emergency management "Plans are useless; planning is priceless." Though the plan was not finished, many elements of Hurricane Pam still proved to be highly useful in response and recovery to Hurricane Katrina days, weeks, and months after the massive storm struck the Gulf Coast.

The IEM Team developed detailed consequence assessments for Hurricane Pam over a scant 53 days. Many of these consequences were eerily echoed in the impact of Hurricane Katrina. Planning needs to be based on sound science of what can be expected during an emergency. I am proud of the work of IEM's scientists and technical professionals and those of our subcontractors in developing a sound foundation for Catastrophic Planning for Southeast Louisiana. Some of the many similarities and dissimilarities are listed below. Please note that the data provided here for Katrina is based on currently available data only. As more information becomes available, some of these numbers could change.

Table 1: Comparisons between projected consequences for Hurricane Pam and actual results produced by Katrina.

"Hurricane Pam" Data	Actual Results from Hurricane Katrina
20 inches of rain	18 inches of rain
City of New Orleans under 10-20 feet of water	Up to 20 feet of flooding in some areas of New Orleans
Overtopping of levees	Levees breached
Over 55,000 in public shelters prior to landfall	Approximately 60,000 people in public shelters prior to landfall
Louisiana Offshore Oil Port (LOOP) shut down pre-landfal and back on in 2-3 days after storm – LOOP handles 12% of US crude oil imports	The LOOP was inoperable from August 29 to September 2 (5 days)
9 refineries shut down during storm	7 refineries in LA shut down during the storm
57 chemical plants shut down during storm. Many flooded and with no power	More than 50 chemical plants shut down during the storm
Over 1.1 million Louisiana residents displaced (500,000 households affected & 230,000 children)	1 million Gulf Coast residents displaced for the long-term; majority are LA residents
Leeville Bridge or LA 1 collapsed (west of city)	New Orleans Twin Span bridge collapsed in sections (east of city)
20,000 boat-based rescue missions and about 1,000 helicopter-based rescue missions	33,500 US Coast Guard missions; 9,313 National Guard missions; 2,911 DoD active duty missions. The Louisiana Department of Wildlife and Fisheries have rescued more than 16,000 people.
786,359 people in Louisiana lose electricity at initial impact	881,400 people in Louisiana reported to be without electricity the day after impact
Over 12.5 million tons of debris	22 million tons of debris in LA; 12 million tons in Orleans Parish; clean up could take up to 2 years
Coastal marsh erosion	Coastal erosion caused by Katrina at landfall equaled one year of erosion in that area (25 square miles a year)
Sewage treatment facilities not working in the metropolitan area	Sewage treatment facilities not working in the metropolitan area
233,986 collapsed buildings	250,000 homes destroyed

"Hurricane Pam" Data	Actual Results from Hurricane Katrina
15% of hospitals in a 13-parish area affected to some degree	All New Orleans medical treatment facilities affected by disaster. 2 weeks post-impact, only 3 out of 29 facilities in Jefferson/Orleans parishes were fully operational; 2 main state hospitals remain closed
$40 billion in damages to LA commercial and residential structures	Costliest US hurricane on record – losses currently estimated at $80 billion
Over 60,000 deaths	1,100 deaths reported to date in Louisiana; over 3,000 still missing
36% evacuated prior to landfall	80-90% evacuated prior to landfall

In the days before and after Hurricane Katrina made landfall, copies of the Hurricane Pam planning document were in great demand. From our review of the response to Hurricane Katrina, it appears that parts of the Hurricane Pam plan may have been used. A federal official guessed that almost 75% of it was used to a greater or lesser degree. Some examples where parts of the Hurricane Pam plan appear to have been used include (note that the data provided here for Katrina is based on currently available data only; as more information becomes available, some of these numbers could change):

- **In Hurricane Pam we projected 36% of the 1.9 million residents and tourists of Southeast Louisiana would evacuate out of the 13-parish region. For Hurricane Katrina, over 80% and perhaps over 90% of residents evacuated out of the region prior to the storm.** Much of this can be ascribed to the accurate scientific forecasts by NOAA and the excellent media exposure of the impending storm. However, the actions of local, state and federal officials prior to the storm were surely in light of the devastating fatalities projected in Hurricane Pam. The city of New Orleans ordered a mandatory evacuation for the first time in its history, at the urging of the State of Louisiana and the National Hurricane Center. The effect of this high evacuation rate is quite clear. The loss of life projected in Hurricane Pam was 61,290. The actual loss of life from Hurricane Katrina in Louisiana thus far is approximately 1,100—although more than 3,000 people remain missing.
- **During Hurricane Pam, participants developed the idea of a "lily-pad" type of search and rescue operation, which was implemented during response to Hurricane Katrina.** Victims were rescued and first transported to a safe area of high ground. From there, another group moved them to land, where they awaited transport to a medical processing center. This allowed Search and Rescue personnel to focus on their primary mission of saving lives. Somewhere between 65,000 and over 100,000 people were rescued during Hurricane Katrina by helicopter and boats. More than 20,000 of these were rescued by the valiant people from the U.S. Coast Guard and more than 16,000 were rescued by the Louisiana Wildlife and Fisheries agency.
- **The Temporary Medical Care section of the plan, specifically the concept of a Temporary Medical Operations Staging Area (TMOSA) was implemented during the response to Hurricane Katrina.** The TMOSAs expanded the triage and

caretaking abilities of special needs shelters reducing the burden on medical facilities. Three TMOSAs were anticipated in Hurricane Pam—Nicholls State University, Southeastern Louisiana University, and Louisiana State University. Two TMOSAs were set up in Hurricane Katrina at Louisiana State University and Nicholls State University. Officially designated as a Search and Rescue Base of Operations, the New Orleans airport effectively functioned as a TMOSA. Almost 100,000 victims of Katrina went through two of these TMOSAs, with more than 15,000 on one busy day.

- **Hurricane Pam planning postulated the use of military bases for staging and temporary housing. After Hurricane Katrina made landfall, many military bases and military vessels were used** including Camp Shelby, Fort McClellan, Lackland AFB, NAS Belle Chase, NAS Meridian, Eglin AFB, Fort Polk, Maxwell AFB, Barksdale AFB, Camp Beauregard, Keesler AFB and NAS Stennis, USS BATAAN, USS IWO JIMA, USS TORTUGA, 14th Combat Support Hospital and USNS COMFORT. These and other installations have housed evacuees in at least 16 states.
- **For Hurricane Pam, it was expected that 1,000 shelters would need to be established and that 55,000 people would be in these shelters prior to landfall. In Hurricane Katrina, responders were able to establish 956 shelters, and approximately 60,000 people were in these shelters prior to landfall.**
- **Hurricane Pam expected that temporarily housing storm evacuees would require 200,000 trailers. After Katrina, 200,000 trailers were ordered for housing evacuees.** Over 140,000 travel trailers and mobile homes are expected to be provided for displaced families, primarily in Louisiana (about 100,000) and Mississippi (about 40,000).
- **IEM estimated that 252,327 children would be displaced from Louisiana schools by Hurricane Pam. During the first Pam workshop, it became evident that there was no plan for schooling displaced children, and plans to address this issue began to be developed.** Hurricane Katrina displaced more than 247,000 public and private school students. The plans developed during Hurricane Pam and afterwards appear to have helped place these children back into school.

When Hurricane Katrina struck, the Hurricane Pam planning was not complete. No training or exercises had occurred using this planning document. The first test was Hurricane Katrina—the deadliest hurricane to strike the United States in recent memory, which would tax even the most mature plans. However, based on the examples presented above, I would venture that, even though the plans and planning were incomplete, Hurricane Pam helped save lives and reduce suffering after the massive catastrophe of Hurricane Katrina.

Recommendations for Preparedness for Catastrophic Events

Managing emergencies is difficult; it has all the complexities of general management with the added pressure of decision making measured not in quarterly returns but in minutes and hours. And unlike in business, the bottom line of emergency management is lives, not dollars. Managing catastrophic events is harder still. Nature and man alike can throw major challenges our way; we need a more mature, robust, and flexible emergency management and homeland security system to respond to these challenges.

From the vantage point of my 25 years working at the nexus of science and technology and the operational world of emergency management, I would like, with all humility, to offer a few recommendations for catastrophe management.

- **Homeland security and emergency management need to be results or outcome-based.** The results from Hurricane Katrina were considered "unacceptable" by the President, by Congress, and by the people of our country. What is acceptable? The political and technical reality is that we cannot have 100% protection 100% of the time. Elected officials, as the representatives of citizens, need to define what outcomes are acceptable. We in Emergency Management and Homeland Security then need to measure these outcomes in our planning, in our training, and most importantly in our exercises. If we cannot produce the level of safety that our nation desires, we need to loop back and see what combination of investments will produce the desired end results.
- **Homeland security and emergency management need to be able to measure levels of safety or protection.** Too often, we look for more and more disastrous scenarios in our planning and exercises without measuring the outcomes that our investments in planning, training, equipment, and exercises have already produced. No business would survive if it could not adequately measure how well or badly it was doing on its bottom line measure. Our nation is spending about $1 million a minute to keep citizens safe. We need to be able to quantify how our efforts are resulting in protection or safety. The tools for making such measurements are available from science and technology; it is time to seriously and comprehensively harness these tools to measure progress. There are myriad ways to improve emergency management capabilities. Some of these can provide large gains in protection from small investments; other well-intentioned strategies can actually *lower* protection levels. We must find ways to measure this protection.
- **A reliable and mature homeland security and emergency management system needs to be developed that can consistently deliver results.** There is, in the human spirit, the desire to see individual heroes. But, it is far more important to develop a system that can allow normal mortals to do their best and produce results with great consistency. To make this happen, emergency management and Homeland Security must become professions with rigorous educational, training and certification requirements. Medicine, law, engineering, warfighting have all benefited from these requirements. So can Emergency Management and Homeland Security. Together with this, Emergency Management and Homeland Security organizations need to undergo appropriate evaluation and certification. All of this requires consistent investments in emergency management.
- **Community planning and development need to be integrated with hazards, threats, and vulnerabilities.** The American economy is vibrant and continues to grow and expand. We need to find a sustainable balance between economic growth and development, ecological and environmental hazards and threats, and the social landscape. There are multiple agencies and organizations engaged in each of these issues and there are few community-based, participatory processes to integrate these together in a meaningful pro-growth, pro-people, scientific manner.

All of these recommendations are feasible and most of them have been accomplished in one community or another. The Best Practices from these serendipitous test-beds can be applied community by community throughout the nation, but especially for regions vulnerable to catastrophic events.

Concluding Remarks

The National Response Plan is a good strategic document. However, integrated Incident Action Plans with sufficient detail are required to handle catastrophic events impacting specific communities. Catastrophes require coordinated action from Federal, State, and local agencies, as well as the private sector. For catastrophic planning to be successful, officials from all levels must be involved and committed to the process and the results. This is not always easy to achieve. There are conflicting priorities, turf issues, and resource concerns. A scenario-based planning exercise like Hurricane Pam makes the disaster real and propels officials at all levels to cut through these concerns and focus on meaningful results.

Hurricane Pam was a step toward this. More than 350 Federal, State, and local personnel—both planners and operational personnel—began tackling the enormous operational complexities involved in responding to catastrophic conditions in an extremely vulnerable area. Historically adversarial relationships were set aside for a few days to work toward the common good—protecting lives and property after a catastrophic hurricane.

Though more workshops to continue the collaboration and planning effort were needed, participants in each group were clearly focused on addressing the catastrophic consequences they had been presented with. Working together, participants developed a mission statement and concept of operations. They also identified response actions to be taken as well as available resources needed to support these actions. They were committed to producing results and there was very little finger-pointing or blame.

Not every region is vulnerable to natural catastrophes, but some are: the San Francisco region, the New Madrid Seismic Zone, the Florida Keys, and of course, New Orleans. For these locations, a detailed and integrated catastrophic plan is the first layer of protection for saving lives.

Hurricane Pam was the beginning of building such a catastrophic event plan. On August 29, 2005, it was at an Alpha stage of release, a version 1.0 of the final. Hurricane Katrina demanded a version 10.0. I urge this committee to consider the value of the Hurricane Pam process and the foundation it offers for other catastrophic plans. From the start, Hurricane Pam was meant to serve as a test-bed for catastrophic planning for other locations. Plans for those locations are yet to be developed. Planning for New Orleans is yet to be finished—with only 127 days left before the start of the 2006 hurricane season.

In August 2005, time simply ran out for one iconic America city. We must not let this happen again.

Post-Hearing Questions for the Record
Submitted to Wayne Fairley
From Senator Tom Coburn

"Preparing for a Catastrophe: The Hurricane Pam Exercise"

January 24, 2006

1. In the wake of a disaster, mismanagement of grants could result in the unnecessary prolonging of the recovery effort. During the Hurricane Pam exercise, how much attention was given to grant management processes and reforms to ensure the grants were being awarded to the greatest need with the greatest efficiency?

Response: The Southeast Louisiana Catastrophic Hurricane Planning Project (SLCHPP) involved workshops that focused on activities related to operational requirements for immediate response activities, such as health and medical, logistics, sheltering and housing, and water and ice To facilitate discussion, a hypothetical hurricane, named Pam, was created and presumed to have agreed upon characteristics in speed and trajectory. Grant management processes were not discussed.

a. Who was identified as the figure responsible for the federal portion of the recovery effort?

Response: The SLCHPP was a scenario-based workshop for the purpose of facilitating the disaster operations planning process, not an exercise. Federal Coordinating Officers and Principal Federal Officials were not designated or role played.

b. Did this person have ultimate authority over how money was being spent?

Response: Again, a Federal Coordinating Officer and Principal Federal Official were not designated and the planning project was not an exercise.

c. Was there a portion of the Hurricane Pam Exercise that identified how contracts for reconstruction and time-sensitive recovery activities could be completed and signed quickly?

Response: Contracting methods and processes were not discussed during the SLCHPP workshop. The planning effort focused on activities related to operational requirements for immediate response activities, such as logistics, sheltering, and the provision of health and medical care, housing, water and ice.

2. For the sake of full transparency and accountability, was there any discussion about

coordinating the creation of a central database and website to publicly list all the contracts awarded during the recovery process?

Response: Development of a centralized database and website to identify contract actions were not subjects of discussion at the SLCHPP. This issue in general was not a part of the SLCHPP.

3. Describe the process that was identified for contract solicitation during the exercise.

Response: This was not an exercise. The SLCHPP workshops were focused on activities related to operational requirements for immediate response activities, such as health and medical, logistics, sheltering and housing, and water and ice. Specific contracting processes and procedures to support response activities were not discussed at the SLCHPP workshops.

a. Please explain any discussions comparing the use of non-competitive and competitive contract.

Response: Specific contracting processes and procedures to support response activities were not discussed at the SLCHPP workshops.

4. During the recovery effort after Hurricane Katrina, FEMA decided to enter into a $236 million no-bid contract with Carnival Cruise Lines to provide additional shelter for evacuees. This was equivalent to 2 to 3 times more than the cost of sending the evacuees on a Caribbean cruise. What processes and procedures were recommended during the Hurricane Pam exercise that could have prevented this waste and abuse of tax dollars?

Response: Specific contracting processes and procedures to support response activities were not discussed at the SLCHPP workshops. The contract with Carnival Cruise Lines to provide additional shelter for evacuees was entered into post-Hurricane Katrina to provide temporary housing. It was not a case of waste or abuse, rather it was a pro-active initiative taken to ensure that evacuees and restoration workers had adequate emergency housing located near the most heavily impacted areas. The contract with Carnival Cruise Lines was the direct result of a solicitation and competitive bid process.

a. Were cruise lines for temporary shelters part of the pre-arranged plan?

Response: Cruise Ship housing was only one aspect of FEMA's broader effort to provide interim and temporary housing to the many people whose lives were disrupted by Hurricane Katrina. The cruise ships provided instant housing for large numbers of people in an otherwise unlivable environment. There was also an immediate need for centrally located housing for first responders - police officers, firefighters and medical specialists.

The cruise ships housed first responders and relief workers close to New Orleans and St. Bernard Parish on short notice where the heaviest recovery work was being done. Excess space on board the ships was filled with local, state and federal relief workers who could not find lodging elsewhere. It would have been impossible for New Orleans and the surrounding areas to progress in their recovery without these workers being able to stay close to where they worked.

5. During the recovery effort after Hurricane Katrina, FEMA decided to enter into a $236 million no-bid contract with Carnival Cruise Lines to provide additional shelter for evacuees. This was equivalent to 2 to 3 times more than the cost of sending the evacuees on a Caribbean cruise. What processes and procedures were recommended during the Hurricane Pam exercise that could have prevented this waste and abuse of tax dollars?

Response: Please refer to our response to question 4 and 4a.

a. Were cruise lines for temporary shelters part of the pre-arranged plan?

Response: Please refer to our response to question 4 and 4a.

6. In the wake of Hurricane Katrina while people in New Orleans were suffering, reports suggest that FEMA spent two days in Atlanta hotels giving classes for their rescue teams on sexual-harassment and the history of FEMA. Meanwhile, Walmart immediately loaded up their trucks with donations of chain saws, work boots, supplies for shelters, and water for the victims.

Furthermore, after Hurricane Katrina and before Hurricane Rita hit landfall, a network of local community and faith-based organizations set up distribution centers across the Texas coastline in anticipation of the imminent disaster. After Rita hit, this network of non-governmental groups were providing relief in parts of Texas that both FEMA and the Red Cross were never equipped to assist. When FEMA and the Red Cross left town, these local groups stayed within their community and continued to offer support to people with long-term recovery needs.

During the Hurricane Pam Exercise, can you describe what roles were given to local and non-governmental entities to help with an efficient and expedited relief and recovery effort? And why did these untrained, local groups have enough sense to adequately prepare for the recovery effort before the Hurricane hit while FEMA did not?

Response: Local and non-governmental entities did participate in the SLCHPP workshops. Participants included but were not limited to the Louisiana VOAD (Voluntary Organizations Active in Disaster), Adventist Community Services, the American Red Cross, and Second Harvest. Participating local and non-governmental representatives focused on the coordination and integration of volunteer support efforts in response and recovery activities. Specific responsibilities were assigned to the Louisiana VOAD and

Adventist Community Services in the Volunteer and Donations Management annex of the draft Southeast Louisiana Catastrophic Hurricane Plan. These responsibilities centered around the coordination of solicited and unsolicited donations and volunteers. Additional responsibilities were not outlined for other local and non-governmental entities. Continued definition of volunteer and donations management under the SLCHPP workshops was cut short by the landfall of Hurricane Katrina.

Post-Hearing Questions for the Record
Submitted to Wayne Fairley
From Senator Daniel K. Akaka

"Preparing for a Catastrophe: The Hurricane Pam Exercise"

January 24, 2006

1. The planners and participants of the Hurricane Pam exercise recognized that damage to New Orleans resulting from a Category 4 or 5 hurricane would leave in excess of 1 million residents unable to return to their homes and over seventy-five percent of structures destroyed. Emergency planners expected that the city would be submerged by water and that structural damage to homes and infrastructure would be catastrophic, leaving residents unable to return to New Orleans for an unknowable, but potentially long period of time.

   What plans were made to provide temporary and long-term housing for such a large number of displaced residents?

   Response: All levels of participants recognized the massive challenges associated with temporary housing in a catastrophic disaster. Participants made progress toward developing a concept of operations to support the establishment of 200,000 temporary housing units, but work was ongoing. Housing options discussed included using existing resources; converting existing resources; constructing emergency group sites; and developing temporary housing sites. Innovative strategies such as the "host city" concept and the use of military bases were outlined as well. Roles and responsibilities for temporary housing were outlined for specific emergency support functions and federal, state, and parish agencies. The planning process had not progressed to fully analyze housing options and develop implementation strategies.

2. Hurricane Pam planners recognized that the communication capacity of first responders and other hurricane relief providers would be severely impacted by the damage resulting

from a Category 4 or 5 hurricane. Planners expected the loss of virtually all land line communications, limited email access, and severely impacted cellular communications.

What steps did the Federal Emergency Management Agency (FEMA) take after the Hurricane Pam exercise to enhance the ability of emergency responders to communicate effectively in disaster stricken areas?

Response: As part of the SLCHPP, 14 workgroups were established to examine specific planning topics. Each workgroup in turn developed communications requirements to be included in a preliminary planning document. The requirements addressed the scope and types of short/long term communications required to satisfy the needs in each planning area, including such things as information sources, information recipients, information content, methods of communications, and technical communications specifications.

This information was intended to be a part of the final version of a planning document that would have included gaps analysis, limiting factors, trigger points for decision-makers, etc. The last workshop was held on August 23-24, 2005, and Hurricane Katrina struck August 29, 2005. Unfortunately, responding to the hurricane required suspension of further work on the Southeast Louisiana Catastrophic Hurricane Planning Project.

FEMA is in the process of assessing, with FEMA Region VI and the State of Louisiana, what the next steps should be relative to combining the information from the Southeast Louisiana Catastrophic Hurricane Planning Project with assessments of the Hurricane Katrina response.

Furthermore, as part of FEMA's Disaster Support Initiatives and retooling, FEMA has undertaken a project to strengthen the readiness of FEMA's emergency communications capability. This project is being undertaken in consideration of the need to have disaster communications capabilities that support response and recovery operations in all situations and particularly when severe damage occurs to the existing communications infrastructure. A three-fold strategy has been adopted to review this issue involving completing a comprehensive review of existing shortfalls; prioritizing shortfalls; and expending funds for execution of prioritized projects. The result will be a prioritized list of projects that will progress toward the goal of "improved disaster communications" allowing more rapid and effective interoperable communications support to disaster field offices and disaster support teams. Finally, this effort is being coordinated with appropriate Federal agencies.

3. During the Hurricane Pam exercise, what was FEMA's expectation as to how long it would take for FEMA to respond to such a disaster?

Response: A specific timeframe was not explicitly discussed during the workshops. As a matter of routine in disaster response, FEMA begins increasing normal readiness in advance of imminent threats that will likely require Federal assistance. This is accomplished by coordinating intergovernmental technical assistance and pre-positioning unique Federal capabilities and assets to ensure timely assistance. All levels of government, Federal, State, and local authorities, perform similar increased readiness activities. Generally as a target, 72 hours is a reasonable timeframe in which to expect disaster response assets to be provided; however, every disaster is different and meeting this timeframe cannot always be guaranteed, depending on the type and severity of the disaster. Some disasters are no notice events. A 72-hour response is more realistic for these events. For those disasters with warning, such as hurricanes, a response time significantly shorter than 72-hours is achievable. As an example, FEMA pre-positioned a record breaking amount of supplies and assets in the Gulf Coast region prior to Hurricane Katrina's landfall. FEMA's heavy push of commodities in the first 6 days after Katrina's landfall provided more trucks of supplies for Katrina victims than were provided for Florida hurricane victims during the entire 7 weeks of the response to their four hurricanes in 2004. Even with pre-positioning supplies, a delay in moving personnel, commodities, and equipment to a disaster area may occur depending on road, weather, and other conditions. FEMA's logistics operations provide comprehensive support and distribution systems for equipment and commodities to effectively support disaster victims and emergency teams responding to disasters.

A major FEMA priority and strategic goal over the next few years is to continue to acquire the equipment and capability to reduce average response times to provide essential services and support to an impacted community with a population of 50,000. As a standard operating procedure, FEMA's logistics operations have attempted to stock supplies adequate to support 10,000 people for a period of 72 hours. Supplies include such things as bottled water, emergency meals, cots, blankets, tents, sleeping bags, etc., and are provided from three large CONUS logistics centers maintained by FEMA. In addition, long lead-time equipment and supplies, such as emergency power generators and plastic sheeting for roofing, are stocked in greater quantities.

4. **Did the Hurricane Pam exercise provide any new insight into what types of assistance FEMA would be expected to provide in the event of a catastrophic hurricane?**

Response: The purpose of the Southeast Louisiana Catastrophic Hurricane Planning Project was to identify immediate operational and planning requirements. The last workshop, attended by State, local, and Federal representatives, was conducted in August 2005. Immediately following this workshop, FEMA was to develop, prioritize, and present formal operational and planning recommendations based on exercise findings to FEMA and DHS management. This process was preempted by the landfall of Hurricane Katrina and the ensuing need for all FEMA and State resources to be applied toward Katrina operations. The recommendations from the Southeast Louisiana Catastrophic Hurricane Planning Project workshops are being analyzed together with the results of FEMA's Hurricane Katrina After Action reviews and recommendations in *The Federal Response to Hurricane Katrina: Lessons Learned.*

The SLCHPP workshops provided greater insights into the strengths and weaknesses of the different participants. Several objectives were achieved during the workshops as follows:

- Review of local, state, and Federal plans (including checklists and decision and action "trigger points"); recommendations of changes needed; and consolidation into a single joint master plan;
- Identification of potentially available existing teams and resources and where, when, and how they will be alerted, activated, and deployed;
- Identification of new plans, checklists, and decision and action points that need to be developed and documented;
- Identification of any inconsistencies or conflicts between State, regional, and national plans and recommendations of changes to resolve discrepancies;
- Objectives with specific emphasis on Search and Rescue, Temporary Medical Care, Sheltering, Temporary Housing, Schools, and Debris; and
- Objectives with specific emphasis on other action topics selected during the exercise, including Hazardous Materials; Unwatering of Levee Enclosed Areas; Billeting of Federal Response Personnel; Donations Management and Volunteer Resources; Distribution of Ice, Water, and Power; Public Information; Transition from Rescue to Temporary Housing; and Reentry and Access Control.

Post-Hearing Questions for the Record
Submitted to Sean R. Fontenot
From Senator Tom Coburn

"Preparing for a Catastrophe: The Hurricane Pam Exercise"

January 24, 2006

1. **Describe the process that was identified for contract solicitation during the exercise.**
   a. **Please explain any discussions comparing the use of non-competitive and competitive contract.**

      - My understanding of your question is this: During the Hurricane Pam exercise, were there any discussions concerning the merits of competitive versus non-competitive contracts for the procurement of response resources for a catastrophic disaster?

        There are several places where contracts are discussed in the Southeast Louisiana Catastrophic Hurricane Plan. These instances mainly deal with situations in which an agency has a standing pre-disaster contract in place that covers certain elements, although there are a couple of instances dealing with situations for which there are no pre-arranged contracts. Due to my responsibilities as the overall coordinator of the exercise, I did not get to spend an extended period of time in each of the rooms that were discussing these issues.

        However, based on my experience in managing emergencies for the State prior to Pam, it was my understanding that any contracts that did not exist before a disaster would go through a contracting office at the agency that was trying to execute the contract—whether that was the Army Corps of Engineers, FEMA, or any other Federal or State agency. It would be up to the particular agency's rules and regulations to determine if the contract had to be competitive versus non-competitive.

        In addition, it had been my experience that if the asset being sought represented a life-sustaining issue, then a non-competitive process was used. However, if it was a resource that was not required to support life, then the regular competitive process was used.

2. **During the recovery effort after Hurricane Katrina, FEMA decided to enter into a $236 million no-bid contract with Carnival Cruise Lines to provide additional shelter for evacuees. This was equivalent to 2 to 3 times more than the cost of sending the evacuees on a Caribbean cruise. What processes and procedures were recommended during the Hurricane Pam exercise that could have prevented this**

waste and abuse of tax dollars?

- The focus during Hurricane Pam was on initially identifying all available resources and means for responding to a potential disaster of a magnitude never experienced before in our country. Refining contracting processes is a very important step, but it was not our focus during this first wave of planning. Presumably, that would have come later.

a. Were cruise lines for temporary shelters part of the pre-arranged plan?

- Cruise ships were part of the Southeast Louisiana Catastrophic Hurricane Plan. On page 87 under the Temporary Housing Section Part 3. Execution, b. Specific Tasks to Lead, Support, and Coordinate Agencies, i. Strategy 1: *Enable use of existing resources, 1) Intermediate housing... Possible alternatives will include college campuses, barracks, hotels and motels, personal travel trailers and recreational vehicles, adopt-a-family, rental rooms in private homes, vacation homes, camp facilities (e.g., church, Boy/Girl Scouts, and 4-H),* ***cruise ships,*** *and all available rental units.*

  During Hurricane Pam, it was always the intention of LOHSEP and FEMA not to exclude any possibilities, even though some, like the use of cruise ships, might have seemed more remote. After Katrina, when no other rental units were available in New Orleans to house the emergency personnel that were so critically needed, using available cruise ships seemed prudent and timely. In addition, this option enabled the emergency personnel to stay together with their families, so that they weren't distracted from their duties by concerns about the welfare of their own families. Also it should be noted that this wasn't just a Federal or State solution; local officials in New Orleans were strong advocates for the use of this resource in the days following Katrina.

3. **In the wake of Hurricane Katrina while people in New Orleans were suffering, reports suggest that FEMA spent two days in Atlanta hotels giving classes for their rescue teams on sexual-harassment and the history of FEMA. Meanwhile, Walmart immediately loaded up their trucks with donations of chain saws, work boots, supplies for shelters, and water for the victims.**

   Furthermore, after Hurricane Katrina and before Hurricane Rita hit landfall, a network of local community and faith-based organizations set up distribution centers across the Texas coastline in anticipation of the imminent disaster. After Rita hit, this network of non-governmental groups were providing relief in parts of Texas that both FEMA and the Red Cross were never equipped to assist. When FEMA and the Red Cross left town, these local groups stayed within their community and continued to offer support to people with long-term recovery needs.

During the Hurricane Pam Exercise, can you describe what roles were given to local and non-governmental entities to help with an efficient and expedited relief and recovery effort? And why did these untrained, local groups have enough sense to adequately prepare for the recovery effort before the Hurricane hit while FEMA did not?

- There are many non-governmental agencies that are very active in disasters. Many of these agencies are active participants in VOAD groups (Voluntary Organizations Active in Disasters). In every disaster that I have worked over the past 15 years, these organizations have been extremely effective because they have **one mission**, helping those who need the help most. Unlike the Federal Government, these agencies don't have all the rules, regulations, and red tape that hinder the Federal response at times.

  These organizations are very vital to the success of any disaster operation and their resources should be leveraged in any disaster. During the Hurricane Pam exercise we invited the Louisiana VOAD to participate, and they sent at least one representative. At the time LOHSEP took the participation of these Voluntary Agencies in disasters very seriously.

Post-Hearing Questions for the Record
Submitted to Madhu Beriwal
From Senator Tom Coburn

"Preparing for a Catastrophe: The Hurricane Pam Exercise"

January 24, 2006

1. Describe the process that was identified for contract solicitation during the exercise.
   a. Please explain any discussions comparing the use of non-competitive and competitive contract.

Contract solicitation was not discussed in the Hurricane Pam workshops, but participants in some rooms did identify areas where contractor support would be needed.

2. During the recovery effort after Hurricane Katrina, FEMA decided to enter into a $236 million no-bid contract with Carnival Cruise Lines to provide additional shelter for evacuees. This was equivalent to 2 to 3 times more than the cost of sending the evacuees on a Caribbean cruise. What processes and procedures were recommended during the Hurricane Pam exercise that could have prevented this waste and abuse of tax dollars?

Budgetary limitations were not directly addressed during the Hurricane Pam workshops.

   a. Were cruise lines for temporary shelters part of the pre-arranged plan?

The plan listed cruise ships as one option for temporary housing. Other options included college campuses, barracks, hotels and motels, and other facilities.

3. In the wake of Hurricane Katrina while people in New Orleans were suffering, reports suggest that FEMA spent two days in Atlanta hotels giving classes for their rescue teams on sexual-harassment and the history of FEMA. Meanwhile, Walmart immediately loaded up their trucks with donations of chain saws, work boots, supplies for shelters, and water for the victims.

Furthermore, after Hurricane Katrina and before Hurricane Rita hit landfall, a network of local community and faith-based organizations set up distribution centers across the Texas coastline in anticipation of the imminent disaster. After Rita hit, this network of non-governmental groups were providing relief in parts of Texas that both FEMA and the Red Cross were never equipped to assist. When FEMA and the Red Cross left town, these local groups stayed within their community and continued to offer support to people with long-term recovery needs.

During the Hurricane Pam Exercise, can you describe what roles were given to local and non-governmental entities to help with an efficient and expedited relief and recovery effort? And why did these untrained, local groups have enough sense to adequately prepare for the recovery effort before the Hurricane hit while FEMA did not?

While coordination with local volunteer organizations was discussed during the Hurricane Pam workshops, the State of Louisiana did not assign specific disaster response and recovery roles to local community or faith-based organizations, other than Adventist Community Services, who were given responsibility to perform management and distribution of donated goods and services. The plan did acknowledge that volunteer organizations often have effective "operational networks" for supporting disaster response and recovery and noted that the State would not impede these operations.

Committee on Homeland Security and Governmental Affairs
EXHIBIT B

Appendix 1: Transportation, Staging, and Distribution Execution Timeline

Pre-Landfall (H-72)

- Federal Operational Staging Areas (FOSA)—Team Arrives (25 people)
  - Unified Command
    - FEMA (1)
    - USFS (18)
    - Operations (1)
    - ESF #5 (1)
    - USACE—Ice PRT (2)
    - USACE—Water PRT (2)
- State Regional Staging Areas (SRSA)—State representative identifies and determines planning items.
  - Planning Items
    - Determine number of Local Distribution Points (LDPs) per parish supported.
      → 15 Type 3 LDPs per SRSA for Day 1 (84 LDPs by Day 3)
    - Determine number of people to start operations.
      → Approximately 45 people/initial operations (~190 full operations)
    - Determine size of footprint.
      → 480,000 ft^2 (10 to 12 acres)
    - Determine number of Material Handling Equipment (MHE).
      → Two full MHE kits (forklifts, pallet jacks, ramps)
    - Determine size of warehouse requirement.
      → 1 to 50 ft. truck of baby food (22/pallet truck) 64 ft^2/pallet—total 1,408 ft^2
      → 1 to 50 ft. truck of diapers (1,408 ft^2)
      → 1 to 50 ft. truck of baby formula (1,408 ft^2)
      → 1 to 50 ft. truck of bottles/nipples (1,408 ft^2)
    - Determine shuttle driver process and additional transportation requirements for specialty items.
      → 25 shuttle drivers/SRSA
      → 26 ft. panel trucks (~14 pallets)
    - Determine ownership transfer.
      → Upon receipt at the staging area

Pre-Landfall (H-70)

- Transport commodities to FOSAs.
  - 18-wheelers (Federal)
- Transport teams and kits to FOSAs.
  - 18-wheelers (Federal)
  - Air (heavy lift rotary wing on request—4)
- Identify/validate sources of buses and drivers.
  - 600 buses—50 passengers each (Local/State/Federal)
  - 1200 drivers—(Local/State/Federal/Volunteers)
- Pre-Stage Louisiana Department of Transportation and Development equipment for ContraFlow.

- Identify personnel and equipment for ContraFlow.

Pre-Landfall (H-50)

- Pre-Stage buses and drivers.
  - 600 buses (Local/State/Federal)
  - 1200 drivers (Local/State/Federal/Volunteers)
- *Phase I Louisiana ContraFlow Plan—Includes areas south of the Intracoastal Waterway. These areas are outside any levee protection system and are vulnerable to Category 1 and 2 storms. During Phase I there are no route restrictions.*

Pre-Landfall (H-48)

- FOSAs
  - Team, equipment, commodities start arriving (150+ trucks)
- SRSA—Pre-Deployment Activities (per staging area)
  - Order Federal resources
    - Shuttle drivers—25+ (if additional are needed, make request)
    - Empty boxes—25+ (if additional are needed, make request)
    - MHE (e.g., forklifts, pallet jacks, ramps)—10 sets (predicted upon breakdown requirement)
    - Personnel—requirements unknown at this time
    - Light Towers—8 to 12 sets
    - Security
  - Fuel and life support requirements for entire staging area team to include contractors, truck drivers, etc.
    - Fuel—1 to 5,000 gallon tank (diesel) MOgas (50/gallon per truck) 24/7 refueling is desirable at the SRSA
    - Food—base camp supplies (contract kitchens)
    - Bathroom facilities—1 porta potty per ~25
    - Contractors (truck drivers, truck support personnel)—power team—100 people
    - Shuttle drivers—25 drivers

Pre-Landfall (H-40)

- *Phase II—Louisiana ContraFlow Plan—40 hours before onset of tropical storm winds. Includes areas south of the Mississippi River, which are levee protected but remain vulnerable to Category 2 or higher storms. During Phase II there are no route restrictions.*

Pre-Landfall (H-30)

- *Phase III—Louisiana ContraFlow Plan—30 hours before onset of tropical storm winds. Includes areas on the East Bank of the Mississippi River in the New Orleans Metropolitan Area which are within the levee protection system but remain vulnerable to a slow-moving Category 3 or any Category 4 or 5 storm. During Phase III, certain routes will be directed and the Contraflow Plan implemented.*

Pre-Landfall (H-24)

- FOSAs/SRSAs
  - Pre-deployment activities continued
    - Finalize transfer of ownership procedures (upon receipt at SRSAs).
    - Government to Government procedures (communication

between Parish, State, and Federal Government)
    - Operational period (24 hrs/day day/night shift)
    - Identify shortages.
    - Identify Communications shortage (e.g., walkie-talkies, radios, cell phones, laptops).
    - Federal partners arrive (State/Federal Coordinators)
        - 2 per shift per staging area
  - State SRSA teams arrive (H-24 to H+24)
    - 190+ people per staging area (working #'s)
    - Receiving—15/shift—30 total
    - Shipping—25/shift—50 total
    - Transload—50/shift—100 total
    - Security—5/shift—10 total
    - Life Support (Medical, Safety, etc.)—TBD
  - Commodities/Equipment arrive

Landfall (H-Hour) ⟵

- Backfill/Sustainment
  - Tractor-trailers (commercial assets controlled by DOT/FEMA/GSA)
  - Buses (commercial assets controlled by DOT/FEMA/GSA)
  - Rail (commercial assets controlled by DOT/FEMA/GSA)
  - Air (commercial assets controlled by DOT/FEMA/GSA)

Post-Landfall (H+24) ⟵

- SRSAs
  - SRSA teams arrive at staging areas
  - MHE arrives
  - Empty trailers/shuttle drivers arrive
  - Commodities start to arrive (H+12 to H+36)
  - Commodities distributed (H+24 continuing)

Post-Landfall – Recon/ Rescue (H+48) ⟵

- Transport responders and commodities to Base of Operations (BOOS), Collection Points, and Temporary Medical Operations Staging Areas (TMOSAs).
  - Air
  - Rail
  - Road
- Transport victims and refugees from BOOS, Collection Points, and TMOSAs to Trans-Shipment Points (Shuttle Staging/Reception Areas).
  - Buses
  - Air
  - Rail
  - Water
- Transport victims, refugees, and commodities from Trans-Shipment Points to temporary and long-term shelters.
  - Buses
  - Air

- Rail

Post-Landfall – LSA Hubbing/ Sustainment

- Transport commodities to Federal and State Staging Areas, Local Distribution Points, TMOSAs, Shelters, Collection Points, Bus Staging/Shuttle Areas, and Trans-Shipment Points
  - Road
  - Rail
  - Air
  - Water

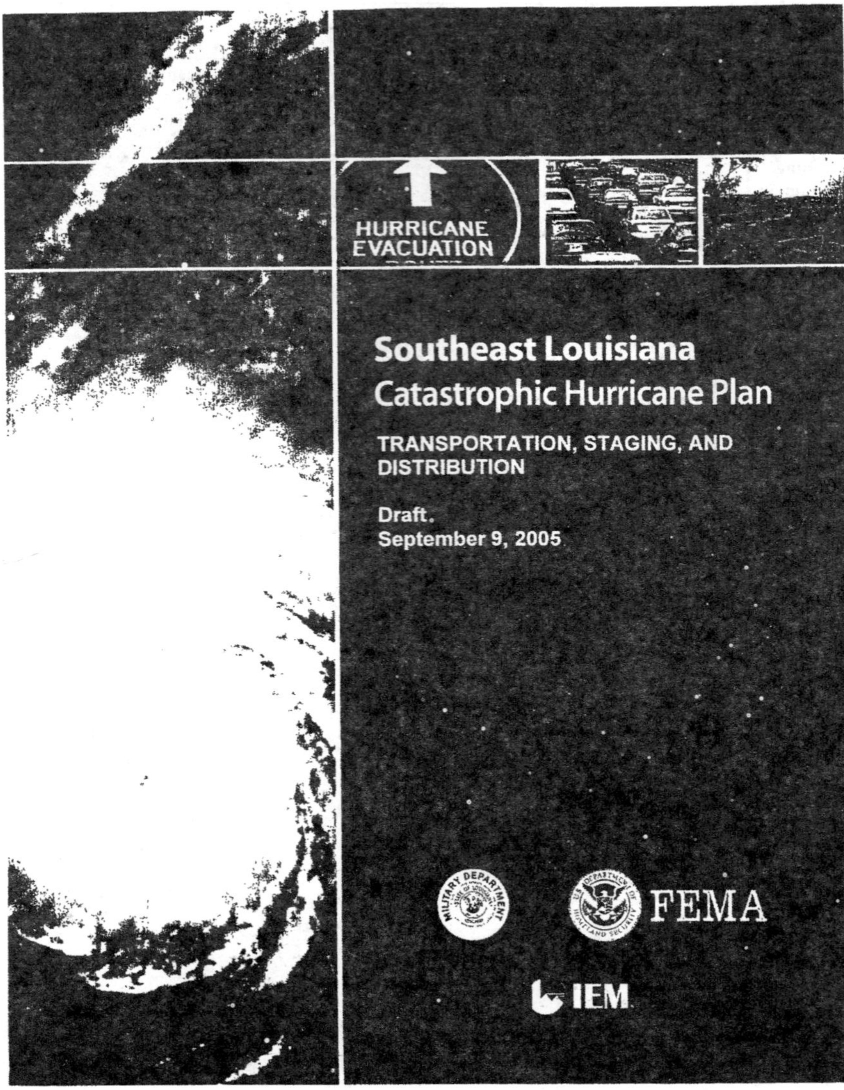
HURRICANE
EVACUATION
Southeast Louisiana
Catastrophic Hurricane Plan
TRANSPORTATION, STAGING, AND DISTRIBUTION
Draft.
September 9, 2005
MILITARY DEPARTMENT
FEMA
IEM

Prepared By

IEM, Inc.
Four United Plaza
8555 United Plaza Blvd., Suite 100
Baton Rouge, LA 70809

Prepared For

DHS/FEMA
500 C Street S.W.
Washington, DC 20472

LOHSEP
7667 Independence Boulevard
Baton Rouge, LA 70806

Prepared Under

FEMA BPA HSFEHQ-04-A-0288, Task Order 001

September 9, 2005

This document is for reference only. Readers should not construe this document as representing official policy or regulations. The functional plan contained in this document was produced during the Southeast Louisiana Catastrophic Hurricane Planning Workshop conducted July 25–29, 2005. The update to this functional plan has not been evaluated for the potential conflicts it may create with other portions of the Southeast Louisiana Catastrophic Hurricane Plan. This functional plan has not been edited from its workshop form. The information contained herein is subject to change.

Appendices to this document are provided under separate cover.

Transportation, Staging, and Distribution

1. **Situation**
   a. **General**
      i. A catastrophic hurricane has made landfall in southeast Louisiana. Its arrival resulted in heavy structural damage due to high winds, a significant storm surge that overtopped levees, and riverine flooding as a result of heavy rainfall. A storm with this magnitude of damage in southeast Louisiana requires efficient and timely transportation, staging, and distribution of critical resources to support emergency response and recovery functions.
   b. **Assumptions**
      i. The affected Parishes will issue a mandatory evacuation order.
      ii. The initial actions of prevention, mitigation, preparedness, response, and recovery operations are conducted by the local government. Local authorities will exhaust their resources, and then use mutual aid agreements with volunteer groups, the private sector, and/or neighboring Parishes.
      iii. The response capabilities and resources of the local jurisdiction (to include mutual aid from surrounding jurisdictions and response support from the State) may be insufficient and quickly overwhelmed. Local emergency personnel who normally respond to incidents may be among those affected and, therefore, unable to perform their duties.
      iv. During a catastrophic incident, the State will take immediate and appropriate action to determine, direct, mobilize, and coordinate resource needs. The State government will suspend and cancel normal operations and redirect resources to saves lives, relieve human suffering, sustain survivors, protect property, and repair essential facilities.
      v. State assistance will supplement local efforts and Federal assistance will supplement State and local efforts when it is clearly demonstrated that it is beyond local and State capability to cope with the catastrophic incident.
      vi. A catastrophic mass casualty/mass evacuation incident triggers a Presidential disaster declaration, immediately or otherwise (National Response Plan [NRP], Catastrophic Incident Annex, December 2004).
      vii. Federal support must be provided in a timely manner to save lives, prevent human suffering, and mitigate severe damage. This may require mobilizing and deploying assets before they are requested via normal NRP protocols.
      viii. A catastrophic incident has unique dimensions/characteristics requiring that response plans/strategies be flexible enough to effectively address emerging needs and requirements.
      ix. Tens of thousands of victims may lack critical life-sustaining resources.
      x. Approximately 10% of the affected population will be self-sustaining.
      xi. In the core area affected by the disaster, it will require more effort to deliver these resources than to evacuate families to shelters. It will be

significantly more difficult to provide logistical support to victims that cannot evacuate.

xii. A catastrophic incident may produce environmental impacts (e.g., persistent chemical, biological, or radiological contamination) that severely challenge the ability and capacity of governments and communities to achieve a timely recovery.

xiii. Sheltering outside the core area may be required for approximately 400,000 to 500,000 people (estimate from the Sheltering Committee).

xiv. Long-term sheltering and temporary housing will be required.

xv. Each person will require one gallon of water and approximately eight pounds of ice per day.

xvi. Patient transportation to and from airheads and medical treatment facilities will be problematic due to excessive congestion on local roads and limited patient movement alternatives (e.g., rotary wing lift).

xvii. Disruption of transportation due to debris and road damage will be a major limiting factor to initial staging and distribution.

xviii. General power restoration will be under the control of the Louisiana State power companies, municipalities, and cooperatives.

xix. Since most emergency generators, and associated switching and control facilities throughout the area are installed at or below the ground level, these generators will not be available for emergency power in the immediate aftermath of flooding.

xx. Since power will be minimal in areas of distribution, availability of fuel will be a limiting factor in determining distribution, including access of victims to the distribution points.

xxi. Competition for refrigerated space for mortuary use may be a limiting factor for ice storage and distribution. Use of refrigerated space for things such as mortuaries and the storage of pharmaceuticals and food will reduce the capacity to provide ice for disaster victims.

xxii. Current staging and distribution sites and transportation routes may be destroyed as a result of wind damage and flooding.

xxiii. The current limited number of staging and distribution sites may result in those sites being overwhelmed by the inappropriate concentration of resources and personnel at each location.

xxiv. The Parishes will request all received materials via the appropriate State channels. No unrequested materials will be distributed to the Parishes.

xxv. Locations, personnel, and supply needs will primarily be addressed at the Parish level during the initial response. Flexibility and adaptation to changing contexts will be a constant requirement.

xxvi. Parishes will have plans in place to respond to the needs of their permanent or "normal" population. Greater adaptation will be required if there is a significant increase in the number of persons in need located within the Parish boundaries and if the resources of one Parish are required in a neighboring Parish.

transportation operations and report promptly as changes occur.

b. **Specific Tasks to Lead, Support, and Coordinate Agencies**

i. Federal Agencies

1) DHS/FEMA will perform the following tasks:

a) Act as liaison to provide logistical support to Federal, State, local, and tribal governments.

b) Set up and stage FOSA sites

i) Supporting Agencies include:

1. U.S. Forest Service
2. U.S. Army Corps of Engineers

c) Working through ESF-1, coordinate the acquisition, movement, and distribution of DHS-owned resources to the incident area.

d) Support ESF-3 infrastructure protection and mitigation missions by providing infrastructure risk and vulnerability assessments in response to actionable intelligence and other information.

e) Activate and convene Federal emergency assets and capabilities to prevent and respond to an Incident of National Significance, and coordinate with State, regional, local, and tribal law enforcement agencies and emergency management organizations.

f) Activate the National Urban Search and Rescue BOOs dependent upon the nature and magnitude of the event.

i) Establish, maintain, and manage the National Urban Search and Rescue Response System. This includes pre-incident activities such as training, equipment purchase, and evaluation of operational readiness.

ii) Manage Urban Search and Rescue task force deployment for employment in and redeployment from the affected area.

iii) Coordinates logistical support for Urban Search and Rescue assets during field operations.

2) U. S. Department of Transportation will perform the following tasks:

a) ESF-1 staff coordinates the acquisition of transportation services to fulfill mission assignments in support of all ESFs when required.

b) The Regional Emergency Transportation Coordinator coordinates with appropriate State, local, and tribal entities to facilitate the movement of people and goods to, from, and within the incident area, and participates in decisions regarding issues such as movement restrictions, critical facilities closures, and evacuations.

c) Coordinates the provision of Federal and civil transportation services in support of Federal, State, local, and tribal entities.
d) Coordinates the recovery, restoration, and safety/security of the transportation infrastructure.
e) Manages the financial aspects of the ESF-1 response, including the funding of Stafford Act mission assignments and/or reimbursable agreements for non-Stafford Act Federal-to-Federal support.
f) Works with State and local transportation departments and industry partners to assess the damage to the transportation infrastructure and analyze the impact of the incident on transportation operations, nationally and regionally, and report promptly as changes occur.
g) Provides transportation support to Urban Search and Rescue task forces and JMTs as requested.
h) Transports refugees to Refugee Collection Points.
i) Transports responders to BOOs and TMOSAs.
j) Transports victims to shelters and trans-shipment points (shuttle staging/reception areas). See Appendix 9: Trans-shipment Points (Shuttle Staging/Reception Areas).
    i) Supporting Agencies include:
        1. Louisiana Department of Transportation and Development
        2. Local Transit Authorities
k) Clear interstate highways, State highways, and bridges of debris that obstructs traffic flow.

3) U. S. Army Corps of Engineers will perform the following tasks:
    a) Provide ice, water, tarps, and operational emergency power at shelters.
    b) Be represented at the Regional Response Coordination Center (RRCC) and the Emergency Response Team–Advance Element (ERT-As)
    c) Provide generators for critical facilities.
4) U.S. General Services Administration will perform the following tasks:
    a) Contracting and leasing facilities
5) U.S. Forest Service will perform the following tasks:
    a) Provide logistical guidance, support, and accountability of commodities at mobilization centers.
    b) Coordinate positive control on trucks of commodities coming in/out of FOSAs by maintaining an information log on trucks received and dispatched from the site (see Communication Requirements).
    c) Provide transportation assets when Forest Service assets are the most effective method.

Committee on Homeland Security and Governmental Affairs
EXHIBIT D

Unified Command Final Briefing
29 July 2005

Temp Housing
Juan Gil
Richard Wiser

State task parishes to provide sites of 20 acres or more—updated annually
See appendix E

Want web-based display/update
SWERN
LONI

Still need to get numbers to go to DOD with to look into military bases

GSA has no inventoried list of sites, but they can research it if there is funding. Needs to be updated annually, too

Will revise appendix E
Need to correct appendix D

Host City concept
Came up with criteria for selecting HCs
Mid-sized cities probably best
Contacting league of cities (?) would help
Need to consider
Schooling, stress on infrastructure

Positive outcomes:
Accomplished preliminary site identification
Web-based database a good idea, excellent tool
Think host city program could be integrated into a nat'l program

QUESTIONS
Jones: was this long-term, strictly for victims?
Ans: we did consider some FEMA people, but mainly focused on the victims from New Orleans.

Did you consider keeping people close to their jobs?
Ans: we want to keep them close, but it may not be possible

How did you arrive at 20 acres?
Juan: that was given by Windell based on what he thought might be available in LA. A 20-acre site will accommodate 160 mobile homes.

200,000 households would require housing (including host city relocation)

Tony: there was no one fix to solve the problem, so they're looking at multiple solutions.

Scott: couldn't hear him. Sounded like a good question about numbers.

Richard: maybe next time we need to look at the numbers to see if we have enough strategies in place to accommodate the requirements.

Scott: talking about making host city a national program. Someone needs to do more than just look at it but work with them.

Richard: yes, there needs to be a lot of research done on this.
We have developed a list of some of the issues that need to be looked at, considered with this (effects on labor pool, infrastructure, schools, med)

Because of the scale of this type of a problem, this is very important.

Where do we go from here on researching this?

I can't hear the answer from the guy in the corner.

Bill: this has been worked on 25 years ago. We should consolidate and organize the effort on this so no one area is doing it unilaterally.

Gary: remember the affected cities are gonna want their people back.

Tony: for some this would be the last resort.

Bill: re the emergency group sites. Governance questions. Have discussions happened with the local authorities?

Richard: we talked about security, postal service…needs more work.

Juan: one of the limiting factors is competing for the limited good places for sites, group sites, sheltering…

Col Smith: how much more work is required to get this operational.

Richard: a whole week with a focus on just numbers

Smith: I want a plan to systematically house at least a big portion rather than waiting til we can totally address everything

Richard: it's operationally ready, needs to be drilled down w/numbers.

Juan: we also want to look into public land, anything belonging to state or private that might be willing to share. We have to be sure we won't be competing with other entities.

Richard: there is a limit to production of mobile homes, etc.

Juan: problems in FL. GSA can deal directly with manufacturers, would help.

Tony: some legislative things might need to change in an event like this. What's our ability to make quick policy changes in the current atmosphere?

Juan: We need to look at how things will be paid for, insurance,

Bill: shopping list of those things would be a good thing to make. There are policies that can be done, you won't have to go back to congress.

TSD
Bill Doran
Kent Weathers

Transportation:
Don Day
Joe Modicut
Jesse St. Amant
Bill Doran

Positive outcomes:
Meeting counterparts on all levels
Long-term refinement needs to continue

We need to further explore the use of the Mississippi River as a via of transport. KEY.

QUESTIONS:
Smith: discussion of getting pres decl. pre-landfall?

Doran: if we don't get it early enough—3 days before landfall—we're way behind. There's reluctance on the part of the state and local to pay for evacuation, etc—they might not have enough.

Smith: this particular circumstance requires a lot of prep.

Doran: we NEED to be able to do pre-landfall declarations

Smith: where is the manpower going to come from?

Doran: we need further planning on that. We're gonna be exhausted for manpower.

Did you discuss EMAC?

Doran: yes, EMAC, Forest Svc., --impressive magnitude.

Tony: let's focus on transportation

Scott: did you work numbers? How long is it gonna take to move people from the water?

Don: process was to bring them into BOOS and TMOSAs, through collection points, quickly to collection areas where there is a transit system to take them to shelters. 600 buses needed just to move people from collection points. The formulas we came up with are more important than the #s themselves. We came up with those. We need to pre-identify the sources for these buses and have them lined up and ready. There are plans to evacuate buses and operators out before the storm. Requires forethought, prior action. We had never looked into what it takes to make a bus staging/dispatch area. That was a great value to these past few days.

We're at less than 10% done with this trans planning when you consider the buses and the people. We've got commodities down.

If you think soup lines in the depression were long, wait til you see the lines at these collection point—and that's the most critical point.

Doran: boats are also important-lilly pads.

Don: We looked at a min of 4 major bus staging areas – geographically focused. Also looked at bus staging areas where buses are already handled (truck stops, schools, etc.).

Doran: buses we are taping will be from the Parish. Might not be able to get a driver. Orleans Parish might be RTA buses. Drivers not provided. Manage evacuation from these bus staging areas.

Joe: DOTD has a unique situation where everybody knows one another and they work very well together.

Jesse: One message to get to FEMA HQ is a pre-landfall declaration in a catastrophic situation is very much needed and should be a requirement. There are a lot of people without personal transportation. Therefore if we don't move people out of Orleans in an appropriate time, there will be mass casualties. The city at this moment does not have the resources or capability to evacuate these people, therefore a pre-landfall declaration is a necessity and a requirement for life safety.
Long term displacement – Long term housing issue is a big logistical problem.
Prioritize: 1) Search and Rescue

KATRINA: THE STORM WE'VE ALWAYS FEARED

The Times-Picayune

50 CENTS 169th year No. 221 TUESDAY, AUGUST 30, 2005 HURRICANE EDITION

CATASTROPHIC

STORM SURGE SWAMPS 9TH WARD, ST. BERNARD

LAKEVIEW LEVEE BREACH THREATENS TO INUNDATE CITY

STAFF PHOTO BY TED JACKSON

NINTH WARD An elderly resident is rescued from chest-high floodwaters by two New Orleans police officers.

By Bruce Nolan
Staff writer

Hurricane Katrina struck metropolitan New Orleans on Monday with a staggering blow, far surpassing Hurricane Betsy, the landmark disaster of an earlier generation. The storm flooded huge swaths of the city, as well as Slidell on the north shore of Lake Pontchartrain, in a process that appeared to be spreading even as night fell.

A powerful storm surge pushed huge waves ahead of the hurricane, flooding much of St. Bernard Parish and New Orleans' Lower 9th Ward, just as Betsy 40 years ago. But this time the flooding was more extensive, spreading upriver as well to cover parts of the Bywater, Marigny and Treme neighborhoods.

As with Betsy, people scrambled into their attics or atop their roofs, pleading for help from the few passers-by.

The powerful Category 4 storm crossed the coast near the mouth of the Pearl River shortly after daybreak with winds of 135 mph. Naval Air Station-Joint Reserve Base in Belle Chasse reported an early morning gust of 105 mph.

With the power out throughout the area and fierce winds raging throughout the day, officials barely began Monday to assess the full damage of the monstrous storm, which was expected to leave thousands homeless and many more coping with damage from the wind and water.

Meantime, ... miles to the west, engineers worked to close a breach along the New Orleans side of the 17th Street Canal.

Huge drainage pumps ordinarily can drive millions of gallons of rainwater uphill through

See KATRINA, A-4

Flooding wipes out two communities

By Brian Thevenot
and Manuel Torres
Staff writers

As Jerry Rayes piloted his boat down St. Claude Avenue, just past the Industrial Canal, the eerie screams that could barely be heard from the roadway grew louder as ... faces of desperate families appeared on ... balconies and in windows, some of them ... white flags.

The scene wouldn't change for the next three hours as Rayes and his ... and nephew boated down St. Claude Avenue and deep into St. Bernard Parish, where water smothered two-story houses, people and animals. The men had to duck to miss ... that towered over Judge Perez Drive, the parish's main thoroughfare.

... Rayes rescued all told the same story, already written on their stunned and shivering

See FLOOD, A-6

INSIDE

PHOTO BY ... / SCO

DOWNTOWN: The damage to the Hyatt Regency on Poydras Street shows that voluntary evacuation is no solution to the dangers of a Category 4 hurricane. See story A-15

After the mighty storm came the rising water

By Doug MacCash
and James O'Byrne
Staff writers

A large section of the vital 17th Street Canal levee, where it connects to the brand new "hurricane proof" Old Hammond Highway bridge, gave way late Monday morning in Bucktown after Katrina's fiercest winds were well north. The breach sent a churning sea of water from Lake Pontchartrain coursing across Lakeview and into Mid-City, Carrollton, Gentilly, City Park and neighborhoods farther south and east.

As night fell on a devastated region, the water was still rising in the city, and nobody was willing to predict when it would stop. After the destruction already apparent in the wake of Katrina, the American Red Cross was mobilizing for what regional officials were calling the largest recovery operation in the organization's history.

Police officers, firefighters and private citizens, hampered by a lack of even rudimentary communication capabilities, continued a desperate and impromptu boat-borne rescue operation across Lakeview well after dark. Coast Guard helicopters with searchlights criss-crossed the skies.

Officers working in the area said virtually every home and business between the 17th Street Canal and the Marconi Canal and between Robert E. Lee Boulevard and City Park Avenue had water in it. Nobody had confirmed any fatalities as a result of the levee breach, but they conceded that hundreds of homes had not been checked.

As the sun set over a still-roiling Lake Pontchartrain, the smoldering ruins of the Southern Yacht Club were still burning, and smoke streamed out over

See BREACH, A-2

INSIDE Residents' insurance claims are to set record A-14 · Looters in the city strike hard and fast A-19 · LATEST STORM COVERAGE › NOLA.COM

30.A1.HURRICANEHITS 2:43 AM

Committee on Homeland Security and Governmental Affairs
EXHIBIT E

Committee on Homeland Security and Governmental Affairs

EXHIBIT F

DHS-FEMA-0055-0002138

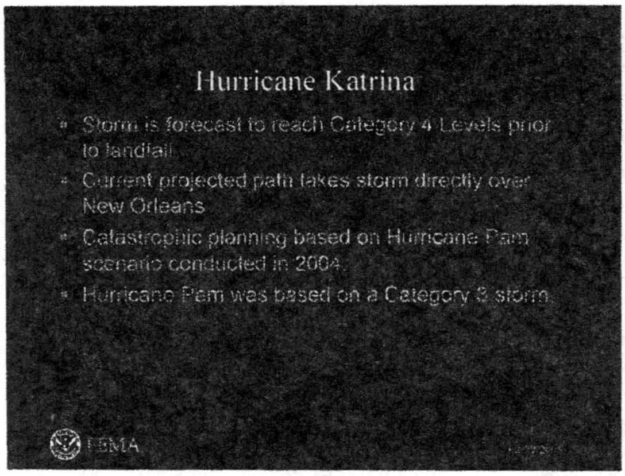

DHS-FEMA-0055-0002139

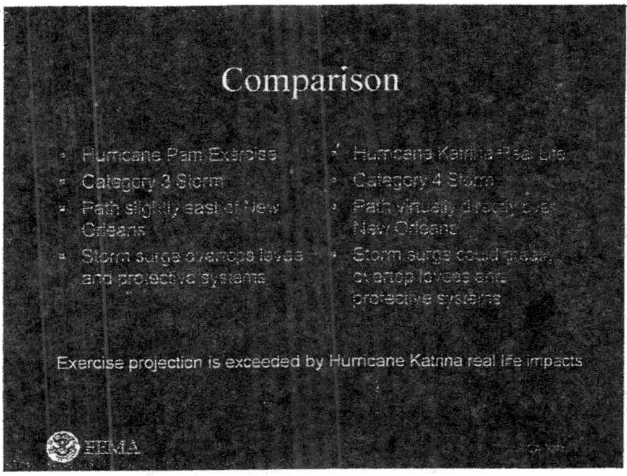
Comparison
Hurricane Pam Exercise
Category 3 Storm
Path slightly east of New Orleans
Storm surge overtops levees and protective systems
Hurricane Katrina Real Life
Category 4 Storm
Path virtually directly over New Orleans
Storm surge could breach, overtop levees and protective systems
Exercise projection is exceeded by Hurricane Katrina real life impacts
FEMA

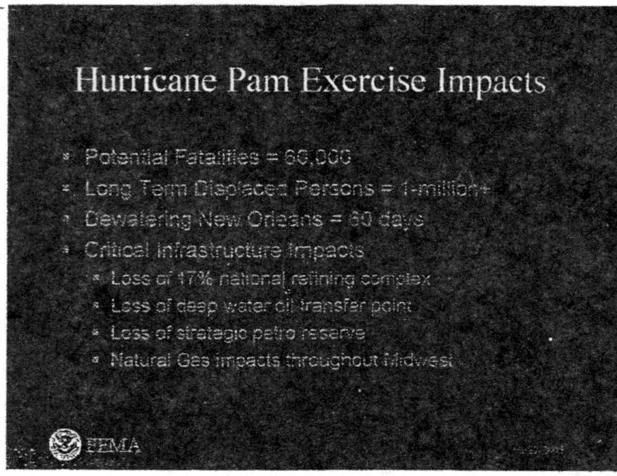
Hurricane Pam Exercise Impacts
Potential Fatalities = 60,000
Long Term Displaced Persons = 1-million+
Dewatering New Orleans = 60 days
Critical Infrastructure Impacts
Loss of 17% national refining complex
Loss of deep water oil transfer point
Loss of strategic petro reserve
Natural Gas impacts throughout Midwest
FEMA

DHS-FEMA-0055-0002141

Impacts

- Long term housing needs – 3-5 years
  - Structures destroyed up to 87% of total
- Loss of Port of New Orleans transfer capabilities
  - Barge and Deep Water transfers
- Incredible search and rescue needs (60,000+)
- Loss of massive infrastructure systems
- Displacement of 1-million+ population
- Unparalleled debris issues (12,606,363 tons)

FEMA

Committee on Homeland Security and Governmental Affairs

EXHIBIT H

New Orleans Hurricane Shelter

Problem:

New Orleans is faced with the reality that it is impossible to conduct a mandatory evacuation in advance of a serious CAT 3, 4 or 5 hurricane. Even under the best conditions evacuation will leave at least 150,000 people in harms way. The City and Federal Government cannot abandon these citizens. Many will lose their lives.

Facts:

1. A Category 3, 4, or 5 hurricane striking directly at New Orleans, or immediately to the west is likely to cost massive loss of life.

2. In a category 3, 4, or 5 hurricane there are no available shelters, provided by Red Cross or any other agency, south of Route 12.

3. Currently there are no authorized city shelters designated for a major hurricane. This is due to the fact that most of the city will be left under 15'-20' of water retained within the levee system. It will take days before outside assistance will be able to "Dewater" the City of New Orleans. Additionally, there is a lack of available structures able to withstand the high sustained winds.

4. Most of the citizens remaining behind after voluntary evacuation are poor, elderly, medically, disabled, homeless and without means (both money and transportation) to leave the city.

5. The City of New Orleans is faced with a choice of developing a "refuge of last resort" or face a potential death toll greater than 9/11.

6. Previously, Federal, State and Local governments have avoided dealing with this problem.

7. The near miss of Ivan and the severe damage caused by this storm has brought the realization that a solution must be developed. Safety and security is the number one responsibility of government.

8. The City of New Orleans does not have the resources to develop a plan or construct an adequate facility. Federal resources must be obtained to resolve this safety issue.

9. Contact with FEMA Director National Preparedness indicates the willingness to work with New Orleans in developing a solution. FEMA understands the potential for massive loss of life if a solution is not forthcoming.

Name of Organization:
City of New Orleans

Project Name/Subcommittees
Emergency Response Shelter/Plan
Homeland Security Appropriations

Organization Contact Details: Terry Ebbert, City of New Orleans, 504-658-6900, tebbert@mayorofno.com.

Public Entity/Non-Profit or For Profit Entity: Public Entity

Project Description/Include Specific Federal Responsibility Funding Project Entails:

The City of New Orleans faces the reality that it is impossible to conduct a mandatory evacuation in advance of a Category 3, 4 or 5 hurricane as well as respond to other natural and manmade disasters, including terrorism. Even under the best conditions that currently exist (in terms of emergency response in the New Orleans region) evacuation will leave at least 150,000 people in harm's way. Currently, no city shelters are designated for a major hurricane or other natural and manmade disasters, and neither the Red Cross nor any other agency provides shelters to address the safety challenges of this region.

Due to its unique topography and location, a serious hurricane event could leave New Orleans under 15 – 20 feet of water retained within the levee system. Unfortunately, New Orleans relies on a system of underground pumps that storm experts predict will take several days or even several weeks to clear the massive amounts of water that could be accumulated. In addition, this unique topography and location makes it especially difficult to address human safety needs for the City's large elderly, poor, medically disabled and homeless community in the event of other natural and manmade disasters, including terrorism. Federal funding is needed to supplement City and State efforts to improve emergency shelter and response plans in the City of New Orleans for these type of natural and manmade disaster events. Funding will be used for final feasibility and planning and design efforts for upgrading the Louisiana Superdome, or any other facility, to serve as a refuge of last resort for major hurricane and other natural and manmade disaster events. The feasibility, planning and design work be used for, but not be limited to: upgrading and elevation of back up power systems with capacity to power all internal systems, structural hardening for Category 3 or higher winds, upgrade of elevated enclosed HVAC systems, upgrade of water and sewer systems to be able to function with loss of city water and power, expanded elevated parking to house larger city emergency vehicles, expanded storage areas to hold stockpiled emergency equipment, supplies, other security, pumping command and control requirements, and addressing the broad range of security and protection needs of inner city shelter areas for civilians in the event of a manmade disaster event, including terrorism.

Requested FY2006 Funding: $850,000

Minimum Funding Necessary/Subsequent Fiscal Year Funding, if any: Any lower amount will allow the project to be started.

Funding History: None—New request

Previous Request(s): None

Administrations Budget Request: None for this specific item.

{W0011472.1}

-2-(Emergency Response, Fy2006)

Requested Bill and Report Language:

"The Committee has provided $850,000 only for the final development and implementation of a major category hurricane, natural and manmade disaster emergency response and shelter plan for the City of New Orleans."

Appropriations Bill/Subcommittee/Accounts/ $ Range of Earmarks in Accounts:

Homeland Security Appropriations, Emergency and Preparedness Response Account, possible options under this main account are:

Minority Emergency Preparedness Demonstration Program;
National Pre-Disaster Mitigation Fund; or
Emergency Food and Shelter Program.

Range of earmarks in these accounts unknown. No specific $ earmarks in previous Homeland Security Appropriations Bills, however, when these accounts and programs were funded in the VA-HUD-FEMA Appropriations Bills (prior to the creation of the Homeland Security Department), earmarks ranged from $50,000 to $1 million.

Authorization Required / Status: None—not required.

Matching Funds: City of New Orleans will provide 10% cash match and/or in kind support for this project.

Funding Agency: Homeland Security.

Supporting Organizations/Including regional/local support: City of New Orleans.

Opposing Organizations: None that are known

Applications for Federal Agency Discretionary or Competitive Grants:

City of New Orleans has solicited funding assistance at the regional level of current Homeland Security Emergency Preparedness offices.

{W0011472.1}

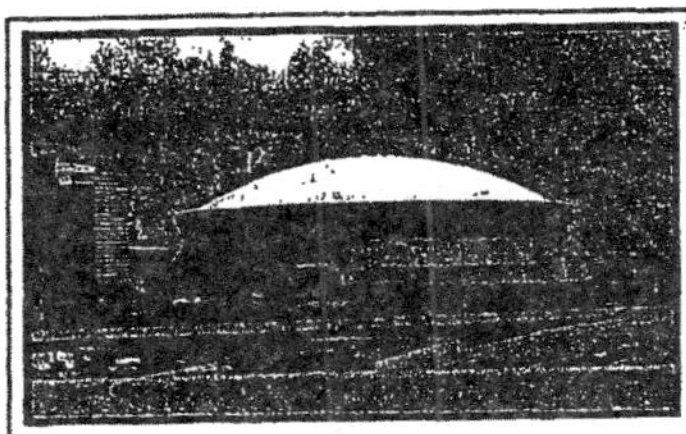

The Louisiana Superdome: "Refuge of Last Resort"

- An extreme event in nature, such as a hurricane, cannot be controlled; but the loss or saving of human life can be dramatically affected by our actions prior to such an event. Every day throughout southeast Louisiana, innumerable Office of Emergency Preparedness staff hours are spent educating the public of the great risk posed by hurricanes and encouraging citizens to establish evacuation plans.

- New Orleans faces some unique challenges: as the only major U.S. city below sea level, our existence is dependent on the levee system and enormous pumping stations. According to experts, a slow-moving category III or any category IV or V hurricane will be catastrophic if it strikes the New Orleans area.

- The safety and well-being of the citizens of New Orleans is the number one priority of Parish officials. Every effort will be made to evacuate the 1.3 million citizens of the metro New Orleans area, but not all citizens may be able to evacuate due to medical infirmity or dependency (oxygen, electricity).

- There are no designated "Hurricane Shelters" in Orleans, Plaquemines, St. Bernard, or Jefferson Parishes due to the inherent risk of flooding in southeast Louisiana. For the more than 57,000 households in New Orleans that do not have access to an automobile, and have not made adequate arrangements for evacuation, this is a life and death situation.

- Under certain circumstances (e.g. greater than 25% probability of a category III hurricane striking New Orleans), the Mayor may mobilize the Louisiana Superdome as a "Refuge of Last Resort." During Hurricane Georges in 1998, approximately 14,000 citizens took refuge in the Louisiana Superdome.

- While the State of Louisiana provides shelters in safe areas outside the region, a significant portion of our population cannot be transported to those sites. Considering New Orleans' immobile population and the substantial risk posed by even a category III hurricane, it is evident the Louisiana Superdome has a critical role to play in the effort to save lives from hurricanes and flooding.

For more information please contact Chief Joseph Matthews, Director, Orleans Parish OEP (504)658-8700

Committee on Homeland Security and Governmental Affairs

EXHIBIT K

Gibbs, Landon M.

From:	Akers, Andrew [Andrew.Akers@dhs.gov]
Sent:	Monday, August 29, 2005 1:47 AM
To:	HSOC SWO, DL-NSC-WHSR, Nichols, Bethany A., Langer, Elliott M., Nielsen, Kirstjen M., Bagnal, Joel B., Farrell, J. Elizabeth, Bentz, Julie A., Kaniewski, Daniel J., Davis, Richard D., Barton, Michael
Cc:	Broderick, Matthew, DiFalco, Frank, Stephan, Bob, Chase, John, Dinanno, Tom, McDonald, Edward, Kulisch, Gail, Paar, Tom, Jackson, Michael (DepSec), Wood, John (COS), NICC, Secretary Briefing Staff, HSOC HSIN
Subject:	Updated NISAC reflecting the CAT 5 Hurricane Katrina
Attachments:	Update of Katrina as Cat 5 for _ 8_29_05.pdf

FYI, Updated NISAC reflecting the CAT 5 Hurricane Katrina

Andy Akers
ASWO HSOC

The Department of Homeland Security's

National Infrastructure Simulation & Analysis Center

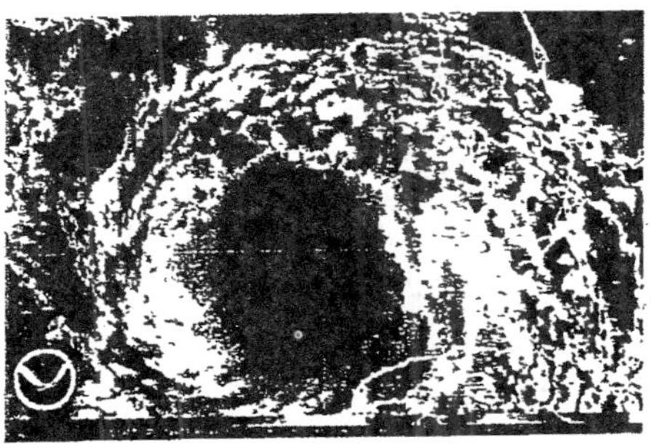

Fast Analysis Report (Update to Reflect Category 5 Status)

to DHS IP on

Hurricane Katrina, Gulf Coast

August 28, 2005

Table of Contents

Executive Summary

On August 27, 2005, the DHS's NISAC commenced an analysis of the potential impacts to infrastructure as Hurricane Katrina moved towards the US gulf coast The trajectory of this hurricane presents the potential for severe damage as it moves through some Gulf Coast oil and gas production and the city of New Orleans.

New Orleans is surrounded by a 130-mile system of levees to protect the urban area, which lies 6 feet below sea level, from surrounding waters. The potential for severe storm surge to overwhelm Lake Ponchartrain levees is the greatest concern for New Orleans Any storm rated Category 4 or greater on the Saffir-Simpson scale will likely lead to severe flooding and/or levee breaching This could leave the New Orleans metro area submerged for weeks or months. At the time this report was initiated, Hurricane Katrina was predicted to reach landfall as a Category 4 storm. (Updates reflecting Katrina's Category 5 status are noted in this updated report) This upgrading to a Category 5 is expected to nominally increase the impacts possibly as much as 30-40% at a minimum unless specifically analyzed. The eye of the storm is forecast to make landfall on the morning of August 29, 2005, and move inland in a north-northeast direction.

Power outage contours were estimated based on the National Hurricane Center forecast as of 9 a.m. MDT (10 a.m CDT, 11 AM EDT) on August 27, 2005 Critical infrastructure elements within the power outage contours have been examined and the results are given here, including an analysis of economic impact.

The following conclusions are supported by detailed information in this document

- Electric power loss is likely to affect over 2.6 million customers, and will be restored over a period of two weeks after damage assessments complete for most of the affected areas excluding New Orleans and the coastal areas Debris generated by the storm will add to delays in restoration of services In total, 133 urban areas lie within the power outage contours; comprising a total population of 2,610.902
- As a Category 5 hurricane, the New Orleans region is estimated to have possible outages that last 16 weeks, if excessive flooding occurs and disables existing pumping stations up to 10 weeks and power repairs that take up to 6 weeks to repair The Coastal area was estimated to have several weeks of business outages due to property damage, power outage, and lack of access to flooded regions. The other areas have expected outage of up to 2 weeks after damage assessments are complete.
- As a Category 4 hurricane, the estimated direct economic impacts of the hurricane are $1.8 to 2.2 billion just for the first week, and estimated total impact to GDP is $7-10 billion for the first week, with up to $3 billion being recoverable due to completed delayed transactions. Continued impacts at lesser magnitudes will continue as the infrastructure is returned to an operating condition. These figures include lost regional economic output and its indirect effect on total output. Louisiana bears the brunt of lost output. This does not include property loss values
- As a Category 5 hurricane, the estimated direct economic impacts of the hurricane are $10 0 to 14.0 billion, and estimated total impact to GDP (gross domestic product) is

$18.1 to $25.0 billion and includes lost regional economic output and its indirect effect on total output Insured and uninsured property losses are not included.

- Separate from the economic impacts is the damage to personal property, which could be as high as $16-20 Billion for a Category 4
- Hurricane Katrina is capable of causing widespread and extended power outages requiring weeks to repair. Restoration may be hampered by flooding or other obstacles.
- A number of generating plants in the likely affected area are coal fired and are supplied by barge, however, the typical coal-fired plant has extensive stockpiles. Moreover, reduced loads (resulting from storm-generated blackouts) will reduce system demand, extending fuel availability windows for generators
- Natural gas and petroleum offshore infrastructure assets have been shuttered in preparation for the storm While the system as a whole has available stored capacity at the refinery level to limit disruption consequences, unexpected occurrences could lead to significant price and demand impacts. Examples include the disruption of refineries in the region which supply 14 % of the nation's refining capability, and of offshore importing capabilities at places like the Louisiana Offshore Oil Port (LOOP) which supplies 10% of the nation's crude oil imports and the majority of supertanker traffic. Even without these disruptions, speculation could lead to uncertainty and volatility in the price of crude oil and refined products in the coming weeks

The baseline analysis in this report is based on the National Hurricane Center forecast as of 11 AM EDT on August 27, 2005, which includes a Category 4 hurricane strike on New Orleans. This magnitude of storm is expected to cause massive flooding, which is not incorporated directly into the outage and restoration estimates, and therefore, is not explicitly incorporated into the economic consequence estimates Some reanalysis has been conducted to consider the upgrading of Katrina to a Category 5 hurricane. Estimates considering this upgrading are specifically noted, all other analysis is for a Category 4 hurricane. Overall the impacts described herein are conservative Due to the unpredictable nature of this event, we are hesitant to predict higher end damage levels, though it is certain that they have the potential to be orders of magnitude greater Any storm rated Category 4 or greater on the Saffir-Simpson scale will likely lead to severe flooding and/or levee breaching, leaving the New Orleans metro area submerged for weeks or months.

The upgrading of the Hurricane strength to a Category 5 also significantly raises the concern and possibility of damage to the energy, petrochemical, chemical and related transportation sector with produces or transports up to 15% of the nation's needs of these products. Significant damage to these systems could have a much longer term impact to the economy that will not be quickly relieved by other capabilities due to a lack of significant surge capacity in these industries (both domestically and abroad). Price increases are expected to occur both due to actual damage and market speculation that will occur nationwide to some extent. The duration of these price increases depends on both the degree of damage suffered, the ability to restore damaged infrastructure, and the ability of the infrastructure and the nation to transparently provide useful information in a timely fashion on the operable status of key assets (such as the LOOP, area refineries and major product pipelines) that minimizes speculation.

1 Introduction

On August 27, 2005, the DHS's NISAC commenced an analysis of the potential impacts to infrastructure as Hurricane Katrina moves towards the US gulf coast. The hurricane predictions were made based on the best data available as of 9 a.m. MDT (10 a.m. CDT) on August 27, 2005. All references to 10:00 a.m. CDT mean 10 00 a.m. CDT on August 27, 2005.

At the time this report was initiated, Hurricane Katrina was predicted to reach landfall as a Category 4 storm. The eye of the storm is forecast to make landfall in area of New Orleans LA, on the morning of August 29, 2005, and move inland on a north-northeast path.

This hurricane follows behind Tropical Storm Cindy and Hurricane Dennis which caused the shut-down and evacuation of some oil-gas production within the Gulf of Mexico. The evacuation of oil-gas production platforms has already begun. The storm has the potential substantially affect the oil and gas industry and the agricultural sector due to the high density of port facilities and agricultural infrastructure around the mouth of the Mississippi River.

2 Storm/Event Data

Figure 1 shows the hurricane trajectory as forecast by National Oceanic and Atmospheric Administration (NOAA) as of 10 a.m. CDT on August 27, 2005. This trajectory and forecast are the basis of this report. The storm track area, shown in white, indicates the range of trajectories possible for this storm. The center of the eye was moving towards the West near 7 mph, and was expected to gradually turn toward the West-Northwest over the next 24 hours. It was a Category 3 hurricane on the Saffir-Simpson hurricane scale, with maximum sustained winds near 115 mph. The hurricane is expected to reach land as a Category 4 storm.

Hurricane-force winds currently extend outward up to 40 miles from its center, and tropical-force winds extend 150 miles. In comparison, the 2004 Hurricane Ivan's hurricane-force winds extended outwards 100 miles, and tropical force winds extended outwards 260 miles.